Mathematics and the Roots of Postmodern Thought

Mathematics and the Roots of Postmodern Thought

Vladimir Tasić

2001

UNIVERSITY PRESS

Oxford   New York
Athens   Auckland   Bangkok   Bogotá   Buenos Aires   Cape Town
Chennai   Dar es Salaam   Delhi   Florence   Hong Kong   Istanbul   Karachi
Kolkata   Kuala Lumpur   Madrid   Melbourne   Mexico City   Mumbai   Nairobi
Paris   São Paulo   Shanghai   Singapore   Taipei   Tokyo   Toronto   Warsaw

and associated companies in
Berlin   Ibadan

Copyright © 2001 by Oxford University Press, Inc.

Published by Oxford University Press, Inc.
198 Madison Avenue, New York, New York 10016

Oxford is a registered trademark of Oxford University Press

All rights reserved. No part of this publication may be reproduced,
stored in a retrieval system, or transmitted, in any form or by any means,
electronic, mechanical, photocopying, recording, or otherwise,
without the prior permission of Oxford University Press.

Library of Congress Cataloging-in-Publication Data
Tasić, Vladimir, 1965–
    Mathematics and the roots of postmodern thought / Vladimir Tasić.
       p. cm.
    Includes bibliographical references and index.
    ISBN-13 978-0-19-513967-9
    1. Mathematics—Philosophy.   2. Postmodernism.   I. Title.
QA8.4 .T35 2001
510'.1—dc21   2001021846

Printed in the United States of America
on acid-free paper

For Maja

ACKNOWLEDGMENTS

As much as I would like to share the responsibility for my oversimplifications, misreadings or misinterpretations with all the people and texts that have influenced my thinking, I must bear that burden alone.

For valuable discussions and critiques, I am indebted to Hart Caplan, Gregory Chaitin, Siniša Crvenković, Guillermo Martínez, Lianne McTavish, Maja Padrov, Shauna Pomerantz, Goran Stanivuković, Marija and Miloš Tasić, Jon Thompson, and Steven Turner. I also thank a number of anonymous reviewers at Oxford University Press, and acknowledge the patient guidance of executive editor Kirk Jensen. Their thoughtful and critical comments led me to rethink the text many times over, and helped me improve it considerably. Last, but certainly not least, I express my deepest love and gratitude to Maja, to whom I am fortunate to be married.

CONTENTS

1. Introduction, 3

2. Around the Cartesian Circuit, 7
   2.1. Imagination, 7
   2.2. Intuition, 10
   2.3. Counting to One, 14

3. Space Oddity and Linguistic Turn, 20

4. Wound of Language, 32
   4.1. Being and Time Continuum, 36
   4.2. Language and Will, 45

5. Beyond the Code, 50
   5.1. Medium of Free Becoming, 53
   5.2. Nonpresence of Identity, 58

6. The Expired Subject, 67
   6.1. Empire of Signs, 67
   6.2. Mechanical Bride, 77

7. The Vanishing Author, 84

8. Say Hello to the Structure Bubble, 100
   8.1. Algebra of Language, 101
   8.2. Functionalism *Chic*, 114

9. Don't Think, Look, 119
    9.1. Interpolating the Self, 123
    9.2. Language Games, 127
    9.3. Thermostats "Я" Us, 132

10. Postmodern Enigmas, 138
    10.1. Unspeakable *Différance*, 139
    10.2. Dysfunctionalism *Chic*, 154

Notes, 159

Select Bibliography, 177

Index, 183

Mathematics and the Roots of Postmodern Thought

1
INTRODUCTION

To know the world, one must construct it.
—Cesare Pavese

The antagonism between two vaguely conceived entities, colloquially labeled "science" and "postmodernism," seems to have become part of public life. In the past few years, pages filled with entertaining invective have sprung up both on the Internet and in print, attracting the attention of such mainstream media as *The New York Times*, *The Guardian*, and *Libération*. These exchanges come under the heading "science wars."

This passionate debate—which revolves around the problem of how science in general and mathematics in particular are read or misread—raises a number of social, historical, and even political questions. I am primarily interested in the following one: *Why* would various postmodern intellectuals bother invoking mathematics in their theories at all?

It could, of course, be a matter of "fashion," as is sometimes claimed. But let us imagine, if only as a counterfactual, that these theorists are trying to convey something that may not be entirely disconnected from mathematics and its history. Can one divine what that is, and how would one go about it?

Whatever it is, it seems unlikely that it would reveal its secret if we restrict ourselves to considering the quasi-mathematical content of assorted postmodern texts. This was fairly clearly demonstrated in Alan Sokal and Jean Bricmont's *Fashionable Nonsense: Postmodern Intellectuals' Abuse of Science* (1998). On the other hand, it could perhaps be recovered by tracing the connections between mathematics and continental philosophy, by searching for historical ties that go deeper than today's tedious incantations of chaos, fractals, and fuzziness.[1]

Perhaps. But there is a small obstacle standing in the way of realizing this idea: No one knows what postmodernism is supposed to be. Attempts to make sense of this elusive concept threaten to outnumber attempts to square the circle. The confusion seems to have reached a peak when the sociologist of science Bruno Latour, who is sometimes categorized as a "postmodernist," published *We Have Never Been Modern*.[2] The term "postmodernism" has become a signifier burdened with so many signifieds that it may well be sinking toward insignificance. (At the least, it appears that its meaning is undecidable.)

It therefore seems best not to rush to judgments and definitions. Rather, I would like to look into the possibility of *re*constructing some aspects of postmodern thought, especially its theoretical aspects known as "poststructuralism" and "deconstruction," from a mathematical point of view. (I will use "postmodernism" as a convenient umbrella term and provide necessary differentiations on a case-by-case basis.)

Mathematics has always been an important testing ground. It is not unreasonable to say, paraphrasing a famous postmodern proverb, that few things can escape the mathematical "text." Mathematics has been part of the Western tradition, inseparable from its culture and its philosophy. Among other things, it has been a source of metaphors. Plato's *Republic*, for instance, advises philosophers that they should study mathematics in order to rise above the world of change and grasp "true being."

The modern era owes a good deal to this old tenet of Plato's. In the sixteenth century, Galileo said that the Book of Nature is written in the language of mathematics. Dutch philosopher Benedict Spinoza authored *Ethics, Demonstrated in Geometrical Order* (1677), whose title, while not self-explanatory, seems telling nevertheless. It has been claimed that the entire project of the Enlightenment had as its goal achieving the clarity of mathematics everywhere by employing the method known as "analytic thinking," whose origins are traceable to mathematics. Even Martin Heidegger, a philosopher known for his sharp critiques of science, endeavored to explain "in what sense the foundation of modern thought and knowledge is essentially mathematical."[3]

It would be possible to skip a few details at this point and tell a wonderfully uplifting story of unstoppable mathematical progress and cultural impact of mathematical ideas—the story of how mathematicians started tackling difficult problems regarding formal reasoning, infinity, sets, logic, and abstract structures; how a number of influential twentieth-century thinkers (e.g., Edmund Husserl, Ludwig Wittgenstein, and Bertrand Russell) were diligent students of mathematics; how mathematics contributed to the conceptualization of computability, intelligence, information, randomness, incompleteness, chaos, and even the conceptualization of the structure of language itself. I intend to tell parts of that narrative in due course. Its importance is impossible to disregard even if it has been told many times.

But mathematics is as much a science as it is an art. It is a peculiar hybrid that—as Byron wrote, not of mathematics but of humanity in general—is "half dust, half deity, alike unfit to sink or soar."[4] Mathematics is practiced by people who are influenced by philosophy and cultural circumstances, by science and poetry, by politics, style, and other passions, by all the traditions to which they belong. Hence, mathematical "accidental tourists" occasionally happen upon places that do not come highly recommended in rationalist guides.

These excursions cannot be ignored here, since my project is to reconstruct certain "antirationalist" exercises of contemporary continental philosophy from a mathematical viewpoint. I would therefore like to consider, among other things, an important part of Western heritage that did not rely on mathematically inspired methods, but that nevertheless informed the views of several influential mathematicians at the turn of the twentieth century.

I have in mind nineteenth-century romanticism. Its philosophical contributions were, for the most part, separate from mathematics and were opposed to the ideal of formal reasoning that mathematics represented. Romanticist rebellion, sometimes called "the counter-enlightenment," is known for its critiques of science and reason. Romanticist "linguistic turn," with its emphasis on the importance of language and culture, art and myth, on the indispensability of imagination and inexhaustibility of the flux of lived experience by means of formal reasoning, played a significant role in placing language and its limits on the philosophical agenda.

In this sense, we might regard the early 1800s as the time when the seeds of the conflict that we now call science wars were planted. Yet the story of science wars seems to be more complex. It is more than a mere episode in a two-centuries-old dispute. It appears, for example, that parts of postmodern theory—despite some superficial similarities with romantic anti-rationalism—find themselves applauding the ultimate in all reductionist projects: artificial intelligence. So the story seems to have at least one unusual plot twist, and we have to proceed slowly, carefully, dusting the web of ideas that veil the mystery of postmodern thought with the care of an archeologist.

When we have glanced at the culture of romanticism—German romanticism in particular—we should be able to put certain mathematical arguments in what I think is the proper context for the discussion of so-called "postmodernism." When, in addition to that, we have examined the cultural influence of mathematical formalism and certain discoveries regarding its limits, we should be in a position to attempt a reconstruction of parts of postmodern argumentation.

I think we will then be able to view postmodern theory as a deeply divided edifice: first, as a revival, or a re-invention in somewhat different terms, of a challenge that mathematicians who were influenced by romanticism once issued to logical reductionism; and second, as an extraordinarily radical dismissal of romantic humanism, a dismissal whose roots can in part be traced to mathematics, and which in its postmodern edition becomes a rather extreme kind of formalism.

To claim that these snapshots suffice to complete the puzzle would, of course, be too ambitious. There are many variations, curious melanges and more or less subtle combinations, of which I can address only a few. Nonetheless, I hope that these notes may offer a representative introductory collage.

I view mathematics here for the most part as a cultural and historical marker, a place where significant cultural events are both reflected and anticipated. But emphasizing the cultural relevance of mathematics—as I have done throughout this text—should not be confused with the gross misstatement that a few mathematicians "have done all that a century ago" and that mathematics is therefore the sufficient cause of this or that event in philosophy.

In particular, only the theoreticians of poststructuralism and deconstruction are in a position to know whether their arguments have been influenced by certain adventures in mathematics. My primary concern is to demonstrate that mathematics *could* have been a formative factor in the rise postmodern theory, and that this possibility stems from the interest in mathematics of its continental "predecessors" and polemical partners. Even on the level of pure possibility, this provides us with a framework for translating parts of the sometimes baffling postmodern rhetoric into a language understandable to the uninitiated, and thus with a context for critically examining various "postmodern" notions.

The sheer quantity of disagreements about the issues at stake indicates that any attempt at telling a truth "beyond reasonable doubt" may end up as an attempt to delude. Therefore, it is probably best to think of this book as a story—a speculative reconstruction of a story—and an invitation to a polemic.

2
AROUND THE CARTESIAN CIRCUIT

How is pure mathematics possible?
—Immanuel Kant

It does not seem reasonable to try to provide a self-contained account of the work of a dozen eminent philosophers in a few pages. Instead of attempting the impossible, I will restrict myself to establishing a few reference points needed for subsequent discussion. There is little content that is specifically mathematical at this stage, because many questions about mathematical knowledge are questions about knowledge in general.

2.1. Imagination

One school of thought, the seventeenth-century rationalism of René Descartes and Gottfried Wilhelm Leibniz, is traditionally believed to have held that mathematics is entirely the business of pure reason. But if mathematics is a mere invention of the spirit, why does it work in practice? If it is reducible to purely logical relations, then it seems that all mathematics is in a sense based on trivialities. However, the wealth and applicability of mathematics make it seem like something that is not at all trivial. We can successfully apply it to what we call the "outside world." Yet Descartes had considerable difficulties establishing even the existence of this "outside world" by purely logical arguments.

Let us leave aside for a moment the well-known theological discourses that Descartes offered to resolve certain difficulties of his approach, and look instead at what he did in practice. As a practitioner of science, Descartes appears to have fancied a dialectical solution of this problem. Perhaps knowledge does not come from pure reason only, in a linear, top-down manner. Something else could be involved, some process of connecting reason with the senses. He seems to have envisioned a kind of feedback loop of mutual justification of theories and facts, where experience and reason cannot be fully separated from one another. I come up with a theory based on observation, and observations in turn justify the theory in a vicious logical circle. This quasi-mystical process, known as the Cartesian Circuit, was for Descartes a proper method.

Descartes, for example, criticized Galileo for introducing the occult hypothesis that all bodies fall with the same acceleration in vacuum. He had no sympathy for philosophers who "neglecting experience, imagine that truth would spring from their brain like Pallas from the head of Zeus." The truth, for Descartes, is neither in the experience nor in the reason alone, but in the circle *itself*. He warns that failing to appreciate that fact and employing methods "pursued by most chemists, many geometers and philosophers" may lead to the occasional success, but that "unregulated inquiries and confused reflections of this kind only confound the natural light and blind our mental powers."[1]

Isaac Newton also had an elaborate version of such a circuit, regarded it as a proper method, and warned against those who do not apply it with due diligence. "I do not feign hypotheses!" he exclaimed. The truth does not belong to experience alone, but it cannot simply spring from one's head, either. Somehow, it is in the circuit itself, although it is difficult to justify that logically.

This might not have bothered Newton too much. The economist John Maynard Keynes collected some of Newton's unpublished notes. He reports that they include rather "unscientific" meditations. Newton was both a master practitioner of science and a thinker with a penchant for the occult: "Newton was not the first of the age of reason. He was the last of the magicians, the last of the Babylonians and Sumerians." He was a "Copernicus and Faustus in one."[2]

Thus it seems that both Newton and Descartes ultimately had to invoke metaphysical powers to guarantee the correctness of their circular methodology.

According to a less mystical school of thought, if (mathematical) knowledge is not based on pure reason or on circular methodology, it seems natural to assume that it comes from the "external" source that is supposedly best understood: sensory experience. This, roughly, is the view of the British empiricists, notably the eighteenth-century Scottish philosopher David Hume. Aware of the difficulties Descartes had encountered in establishing the existence of the external world by rationalist arguments, Hume took a more skeptical stance and maintained that philosophy cannot go beyond experience.

He held that all knowledge can be seen as "relations of ideas" and "matters of fact," but that all logical concepts and relations, such as causality, are acquired from experience.

But in this case, how can we be certain that mathematical statements are true? Knowledge acquired by collecting sensory experiences cannot be guaranteed to be true in all future instances to which it is supposed to apply. Yet mathematics appears to be true forever, in the sense that it does not require subsequent experimental confirmation.

Hume's skepticism thus appears to question the universality and necessity of mathematical knowledge. This does not mean, as it often seems to be assumed, that Hume outright denied the necessity of mathematical truth. In fact, he states that despite our inability to justify our knowledge, we continue to act *as if* we were in the possession of it. He is skeptical about rational justifications of knowledge, but he cannot deny that we believe we have it. This strange belief, he says, we get by employing a somewhat vaguely conceived faculty of imagination. He more or less leaves it at that, but it is important to note that he does not take much pleasure in his own skeptical conclusion and hence posits imagination as a sort of intermediary, however faulty and unjustifiable, between reason and the senses.

This was the situation until relatively late in the eighteenth century. Various people seem to have entertained the possibility that there is something extra-logical about mathematical knowledge, even if they came to that point from different premises. Descartes toyed with the mysticism of his viciously circular method. Hume maintained cautious skepticism, but played with the idea of imagination, a "fiction" by which nature itself saves us from the insufficiency of pure reason.

Somewhere between Descartes and Hume, in time and in outlook, was the eighteenth-century Italian philosopher Giambattista Vico. Slightly before Hume, and in a different way, he expressed the idea that truth is somehow "imagined" or "made." However, Vico's fundamental principle—"the truth is the made"—would become known to a wider readership only in the nineteenth and twentieth centuries.

But in what sense is the truth supposed to be *made?* It was partly for the purposes of explaining this that Vico introduced his own concept of imagination (*fantasia*). Like Hume, he maintained that we cannot have proper knowledge of natural laws, not because Vico was a skeptic but because we do not make these laws. We can only have a kind of "consciousness" *of* them, because they come to us by means of sense-experience. Unlike Hume, and closer to Descartes, whose metaphysics he nevertheless attacked, Vico thought that we can have the knowledge of mathematics, which is "universally imagined." Mathematics is a reflection of the general form of our experiences, all of which are governed by imaginative universals (*universali fantasia*). This is an important idea. If our experiences always have a particular form, as represented in mathematics, then every experience, in a "Cartesian" circularity, must reconfirm our mathematical knowledge. So the circularity becomes "unnecessary," although it very quickly appears elsewhere.

To see where the circuitry resurfaces, we must ask Vico to explain exactly where our universal imagination comes from. For Vico, it is a matter of history. This is where the circle now resides. History always moves in cycles (*corsi e ricorsi*), moving from the age of gods to the age of heroes to the age of humans, then back. In the first two stages, the universal poetic imagination governs all thought and orders all experience. In the age of humans, this imaginative sense is lost, and all thought becomes abstract and ineffective. Then the cycle must begin again.

The Cartesian Circuit, in one form or another, seems to be the specter that haunts a good deal of Western thought. The problem of "squaring" this circle, ironing it out, displacing it, justifying it, or accepting it as it is, is a marker that I follow in the rest of this chapter.

2.2. Intuition

Late in the eighteenth century, German philosopher Immanuel Kant presented an influential attempt to formulate a theory of knowledge, including a philosophy of mathematics. Kant's philosophy can be seen as an attempt to investigate the somewhat unclear connection between reason and experience. Descartes's mysticism probably seemed to him as unsatisfactory as Hume's skepticism. (It is not clear whether he knew about Vico.)

Kant's motivation to reconcile these views could be described as follows: Concepts without experiences are empty; experiences without concepts cannot constitute knowledge. In the *Critique of Pure Reason* (1781), Kant says:

> From what source do we derive [our] concepts? If we derived them from the object [...], our concepts would be merely empirical [...]. And if we derived them from the self, that [...] could not be a ground why a thing should exist characterized by that which we have in our thought [...].[3]

It was in the search for the source of our concepts and their connection to experience that Kant proposed to change the entire framework of understanding. Kant is in this respect a crucial as well as controversial figure of Western philosophy. Albert Einstein once said that every philosopher has his own Kant. For all its importance and influence, it seems best to regard Kant's project not as an attempt at ultimate philosophical grounding, but as a revolutionary proposal for reconsidering our philosophical outlook.

Both empiricism and rationalism in a sense made the external objects central and attempted to describe how the mind can come to know them. We have seen that this leads to some difficulties. Kant's idea was to make consciousness central. This change of viewpoint is known as Kant's "Copernican revolution." It is not that I am conscious of an object *itself*. The object is for me always the object *of* my consciousness; it is a representation. Since Kant does not deny that the thing *is* out there, it follows—Kant argues the point in some detail—that the thing-as-I-know-it (*phenomenon*) cannot be said to be the same as the thing-in-itself (*noumenon*).

For example, the desk in front of me could be described both as a collection of molecules and as a brown wooden object with four legs. The table, in itself, fits both descriptions, although I usually do not describe it as an aggregate of molecules. In this sense, my descriptions do not exhaust what the desk *is* in itself. From that standpoint, a science such as physics does not offer direct knowledge of the physical universe, but rather conceptual descriptions of the observer's experiences of being in the universe. Since we seem to have our own viewpoints and different personal experiences, the question arises of whether there is something universal and necessary about our descriptions, something that we could call "knowledge."

Kant sets for himself the task of determining the conditions under which knowledge is possible in his "revolutionary" framework. It is not necessary to get into the intricacies of his arguments. At the moment, I am interested only in the idea that there is something extra-logical that is the *precondition* of all knowledge. There is something that governs the form of all experiences and all thought, something like Vico's imaginative universals, which synthesizes concepts and experiences into knowledge. The comparison of Kant's notion of a *schema*—"representation of a universal procedure of imagination in providing an image of a concept"—with Vico's *fantasia* is not entirely accurate, but Kant's objective seems to be roughly analogous to Vico's. If we know that our experiences always take a particular form, then we are a step closer to ironing out the Cartesian Circuit.

If I come up with a conceptualization of the form of my experiences—for instance, a mathematical conceptualization—and if I know that my experiences are of that form, then it appears that no further "experiment" could ever refute my theory (at least if I manage to demonstrate that my concepts correspond to *possible* experiences). According to Kant, the form of experience is given by the *a priori intuition of space* and the *a priori intuition of time*. Without these intuitions, I would not even be able to distinguish myself from other things. The intuition of space would provide me with my "outer sense," while my "inner sense" would correspond to the intuition of time.

Consider the problem of counting objects. Since I can experience an object, and then experience it *again*, it follows that I can apply myself to distinguish between (the two) experiences: One is subsequent to the other. This in turn seems to imply that my ability to count involves the a priori intuition of time. Suppose that I experience some object now, and then again seconds later. Say there is an apple in front of me. I see an apple, I close my eyes; then I open my eyes and see an apple again. I had two experiences, but how many apples have I experienced? One or two? Time alone tells me nothing about that. Thus, to count objects properly, I need something more than just an intuition of time. I need the ability to identify objects as being the same, in time. If I am Kant, I will not call the apple the same unless I see it at the same location. For that, however, I need a sense of space.

Both of these intuitions function as molds in which our experiences are shaped. Based on the *form* of our experiences, we come up with a conceptual framework that describes them. For example, geometry and arithmetic are such conceptual frameworks. Thus, it seems that statements of geometry and arithmetic can never be empirically falsified. They are based on the very form of representation of all sense-experience. This would then account for the apparent necessity, or truth, of mathematical propositions. We cannot have an experience of them not being true, any more than we can have an experience of the same physical object being at different places at the same time.

This at best explains our *belief* that mathematics is true. But belief is not the same as knowledge. I can easily come up with a concept such as "round square," and I may even convince myself that round squares are round. However, this is an empty concept. The intuition that led me to form the concepts "round" and "square" prevents me from attaching the description "round square" to any object of my experience. It is not possible for me to experience anything that is both round and square, so I cannot claim that my statement that "round squares are round" is knowledge.

Even if a statement such as "round squares are round" is apparently tautological, Kant does not consider it knowledge. If there were something that were both round and square, then it would obviously also be round. Unfortunately, it would also be not-round, because it would be square. Both statements seem true, but they contradict each other. This is of course a trivial example, and it does not cause difficulties. But it points out that dealing with empty concepts can be deceiving. Pure reason sometimes represents two incompatible statements as formally true. That is for Kant one of the fundamental deceptions of pure reason, which he calls "antinomies" (the simplicity of the above example notwithstanding).

Kant's antinomies are concerned with deeper instances of such deceptions, in particular, the nature of space and time. I discuss some of them in due course. For now, it is important to note that knowledge, for Kant, must correspond to a possible experience. What seems like a logical possibility could end up as an empty concept and thus lead to incompatible assertions similar to those about the round square. Reason is reliable when it deals with phenomena, but not when it deals with (possibly empty) ideas or with things-in-themselves. Therefore, an explanation of how mathematical knowledge is possible requires something more than just having the intuitions of space and time. It requires our ability to know that mathematical concepts correspond to possible experiences.

Part of Kant's theory seems aimed at providing a methodology for deciding *which* pure concepts can be seen as possible experiences rather than as mere logical possibilities. Thus, to provide a "Kantian" grounding of mathematics as knowledge, it does not suffice to demonstrate that mathematics is *logically* possible, namely, that it is free of contradictions (which itself turned out to be difficult enough, as Kurt Gödel and others showed in the 1930s).

Instead of mere logical possibility, we should have something like a "real possibility." But classical logic—or "general logic," as Kant calls it—cannot deliver that type of argument. Even Leibniz allowed the leap from logical possibility to real possibility only with respect to his proof of the existence of God: Only God, says Leibniz in *Monadology* (1714), has the privilege of necessarily existing because He is possible. Kant was concerned about this. He comments in the *Critique of Pure Reason* that "the celebrated Leibniz is far from having succeeded in what he plumed himself on achieving."[4]

While the application of "general logic" within mathematics is not called into question by Kant—an attempt at such a coup d'etat had to wait until the early 1900s—its applicability on a "meta-mathematical level," the level of "transcendental philosophy" at which the possibility of mathematical knowledge is to be established, is at best limited. For this reason, Kant proposes a somewhat different kind of logic, which he called "transcendental":

> General logic [. . .] abstracts from all content of knowledge [. . .]; that is, it treats the form of thought in general. But [. . .] we should have another logic in which we do not abstract from the entire content of knowledge. This other logic [. . .] should contain solely the rules of the pure thought of an object.[5]

Consideration of such a logic, and the problem of constructing a meta-mathematical argument that would establish the possibility of mathematical knowledge, makes Kant an important (if sometimes implicit) dialogical partner in the early-twentieth-century debate about the foundations of mathematics.

There are, however, some difficulties involved in Kant's approach. Some of them are related to certain discoveries in geometry, but I leave that aside until chapter 3. The problem is: How can I *know* that my experiences are governed by intuition of a particular kind? The question in turn leads to a more general problem about Kant's viewpoint, and seems to have been the cause of some disagreements among nineteenth-century philosophers.

2.3. Counting to One

Kant claims to have deduced that the intuitions of time and space are a priori given to the mind. Many people would agree that the human mind has certain innate capacities. For instance, it has been argued that the capacity to learn a language is innate. Kant did not have much to say about language, and some of his contemporaries were quick to reprimand him for that. However, from his descriptions of just what is given to us about space—such descriptions appear in some of Kant's statements about geometry, and we will glance at them in chapter 3—it appears that he believed it is possible to ascertain the precise manner in which intuition shapes our experiences. It is reasonable to ask how we could *know* that this is so, for instance, based on Kantian rather strict criteria as to what constitutes knowledge.

One might say that a certain amount of reflective, introspective activity suffices to determine the precise nature of what is intuitively given to the self. Applying Kantian criteria for sifting out actual knowledge from the "empty," formal statements of general logic, it would seem to follow that such introspective activity can lead to knowledge only if it corresponds to an experience of something, a phenomenon, namely, myself. In other words, it appears that I would have to become an object of my knowledge. This creates difficulties.

To show what the difficulties are, let me ask what may seem like an extraordinarily silly question: If I am a knowable object of my own experience, then how many of myself *are* there? I would not be too happy if the answer turned out to be, say, forty-two. What I would like is to have the knowledge that there is only one of me out there, at least in the realm of the objects of *my* experiences.

Now recall how Kant counts objects. To begin with, these objects must appear to me as being in space and time. I could have indefinitely many experiences of an apple as an object of my thought, but it is only upon its re-identification in time *and* space that I can establish its uniqueness as an object. The apple has a certain permanence, an independence that *affects me* in some way, which then enables me to grant it the status of objecthood. Something is an object if it is in space and time and is relatively inert with respect to myself.

This involvement of my self with the object short-circuits when I try to conceive of the I as an object. It appears that I can only know myself as something in space and time, something that is relatively inert with respect to me, and even affects me in some manner. So I would have to be relatively inert with respect to myself, which seems a tad strange. Kant was aware of this problem: "We intuit ourselves only as we are inwardly *affected*, and this would seem to be contradictory, since we should then have to be in a passive relation to ourselves."[6] Hence, I can conceive of the possibility of "counting myself as one" only by mobilizing my outer sense, that is, by invoking something *other* than myself. But then I am somewhat removed from *intro*spection.

Kant "resolves" this by saying that I have an awareness of the self, a presence that can be sensed, but he does not call it knowledge. The unity of this awareness, my thought of the self, is a *formal* unity. It is not knowledge (of something)—it is more like a conviction, a certitude that is a precondition of all knowledge.

I can, of course, try to give a detailed taxonomy of what this "formal unity" provides for me—that is, try to determine the preconditions of my knowledge—just as I can try to classify infinite sets, formally, despite the fact that I cannot have much experience of these things as objects. Kant did attempt such a description of various aspects of the human condition and its epistemological preconditions. These descriptions would likely be unreliable because they are based on introspection, and the beam of introspective insight, as I indicated above, cannot lock onto its target object. Something larger than myself, other than myself, seems to be involved. (Kant's insights into the precise nature of our intuition of space soon came under attack from mathematicians.)

It therefore seems that I need something more than introspection to support my certitude about things. Kant seems to have been aware of that:

> [A] reflective and enquiring being should devote a certain amount of time to the examination of his own reason, entirely divesting himself of all partiality and openly submitting his observations to the judgment of others [. . .], that is, before a jury of *fallible* men.[7]

Kant is apparently saying that no one is entirely infallible. We are autonomous individuals, we have some intuitive, introspective insight, but we are fundamentally involved in the community to which we submit our observations in the hope of remedying our individual limitations. Yet it is not possible to reduce this interaction to a simple formula. Community seems to have a role, but Kant would have regarded individuals subscribing to community's beliefs without critical examination as "immature" and "unenlightened."

Whatever Kant himself meant, the point was soon taken up by several of Kant's contemporaries. First consider Johann Gottlieb Fichte, an important German thinker whose ideas were central to early romanticist philosophy. We saw that Kant faced the problem that in order to know oneself, to "count oneself as one," one must invoke something other than oneself. Fichte makes this into a kind of principle: Without the not-I, there can be no I.[8] (It is a special case of a more general principle: "Every determination is a negation." Determination of something involves distinguishing it from other things.)

This seems fair enough. It would be a tedious world if it consisted of me only. So I must know something other than myself, something that I am not.

But how can I know what I am not, if I do not already know what I am? What, precisely, should I be looking for?

Fichte deals with this in the following way. My being has some initial familiarity with itself. However, I can only advance from this initial point by interacting with things other than me. I can reflect upon myself all I like, but, as the German poet Novalis wrote, "what reflection finds, seems to have been there already."[9] Introspection is not enough, so I am always in the need of some "other," something that I am not. And I can only know something that I am not by noting its "impact" on me, how it affects me, its resistance to my actions. Hence, I must always *act*, realize myself in the larger, resistant world of the "not-I": material world, language, culture.

For Fichte, the self is the knowing-I, the ego, *and* something else in addition to that, a creative force that I can *conceive of* as a kind of continuous action although I can never *know* it. There is "something more in me than me." The assertion is often heard in our postmodern times—it sounds like a confusing psychoanalytical mantra—but the idea goes back to romanticism and basically means that there is something in me that is beyond my objective knowledge. This leads to the talk of the "spirit," of "self-consciousnes in general," or "absolute ego," which in turn causes other difficulties. What is the absolute ego, this "self-consciousness in general"? Whatever it is, it precedes all knowledge, and it is in some sense beyond language. It is not even an "it," because it is not a static presence or a thing. It can only be conceived as a flux, a will, a drive to act. "We do not act because we know," says Fichte; "we know because we are called upon to act."[10] Or, in the words of Goethe: "In the beginning there was the act."[11]

What remains in terms of cognitive activities is the endless cycle of creative self-realization in nature, culture, art, and language, and the feedback resistance of these not-Is to my effort, the resistance that in turn lets me "know" myself as a continuous action. I can employ my imagination to describe this process, but no description is completely adequate. My imaginative constructs are historically conditioned and cannot be subdued to a forever-fixed classification. (Einstein, who along with several mathematicians imported some romanticist ideas into the world of science, expressed a similar opinion. There is something that bridges the gap between experiences and concepts—we can call it "intuition" or "imagination"—but it cannot fit an unalterable categorization.)

The idea of a continuous "creative" activity that cannot be captured in terms of an a priori fixed language would exert some influence on a number of mathematicians later. I address the effect of this notion on mathematics in chapter 4. What is important for now is that for many romanticist thinkers the incompleteness of knowledge becomes quite natural. Indeed, it is one of the fundamental principles of romanticism—if there is such a thing at all—that no knowledge can reach this irreducible "active ingredient" of the self.

It can be sensed, felt, lived, but neither known as an object of scientific study, nor *should* science be able to reach this primary force of nature. If it did, if science and logic and language could capture it in some way, there would be no freedom to act creatively and—so romanticism argues—everything would turn into a rigid deterministic scheme. This sense of dependence on the inexplicable is for romanticism of the root of art and ultimately of all religious feelings. God is no geometer, rather an unpredictable poet. (Geometers can be unpredictable poets, so there could be room for compromise.)

Let us also glance at some ideas of another important German philosopher of the same era, Georg Wilhelm Friedrich Hegel. Hegel worked among various romanticist thinkers, but he developed their ideas in a different direction. He was not content with saying, as some romanticists did, that the process of artistic creation is simultaneously a philosophical paradigm, a way of life, and something that is ultimately beyond "scientific" descriptions. The limits of individual knowledge, set out by Kant and radicalized by romanticism, are for Hegel the limits of individuals and not of knowledge.

For this reason, Hegel takes the point of view of knowledge itself and makes it into a subject that he calls "the spirit." Unlike the romanticist absolute ego, which can only be known as an endless striving, Hegel argues that the spirit indeed reaches a state of self-knowledge. This seems a little occult, so imagine instead that "the spirit" is something common to all individuals, some mysterious "thing" that, just like mathematics and law and the Dow Jones industrial average, is a historically conditioned social "fiction" that can nevertheless have nonfictional consequences for individuals (something like Vico's imaginative universals, except that it seems to have a mind of its own).

"What is rational is real," says Hegel; "what is real is rational."[12] This is a famous statement that can be (and has been) interpreted in a variety of different ways. Let us approach it indirectly. We always believe that our knowledge is real—in the colloquial sense of allowing us to rationalize reality—so, conversely, our reality can be regarded as rational. However, some new piece of reality might turn up, because everything is in a fiery Heraclitean flux, and it will have to be *made* rational. The new rational explanation might annihilate our previously "real" knowledge, which must then be reformulated to rationalize the evolved actuality. This seems like a distant relative of the Cartesian Circuit, but to understand Hegel properly we should try to do away with the dualism of knowledge and reality involved in my slightly deceiving analogy.

But how can this circularity be described without reference to evolving realities to which knowledge, as it were, perpetually adjusts? It might be difficult, but it is not inconceivable. Mathematics certainly experiences changes due to the evolution of its own "realities," although these "realities" are far from what most (real) people would regard as real. The theory of computability, for example, has had a profound impact on mathematics as a whole. Even more interesting, it is now routinely invoked in linguistics and cognitive science. The abstract notions of a few mathematicians and logicians who were interested in the conceptualization of computation and its physical realizations—theoretical models of what we now call "computers"—have affected society in ways that we have yet to understand.

So knowledge systems evolve and change, and not only due to the discovery of curious weather patterns that people would like to explain. It therefore seems unsurprising that Hegel attempted to give an account of such changes "from the inside," from the point of view of the spirit. Perhaps inspired by Kant's antinomies—where pure reason misleads by presenting incompatible statements as being "true"—he notes that every concept, by virtue of being reflected upon, necessarily invokes its negative, just as (already for Heraclitus) "large" cannot be thought of without "small."

This opposition, when reflected upon, itself produces its negative, which takes the form of a new unity of the original opposites on a higher reflective level. Thus, the original concept is "annihilated," but at the same time preserved and "elevated" to the higher level of reflection. It is because of this preservation that the process is supposed to have the good character of a circle or a helix, as opposed to the bad character of "running in circles."

Something must be preserved because annihilation is itself unthinkable without its negative. What is preserved, loosely speaking, is the memory of this act of negation of the previous conceptual framework. Hence, the old framework continues its life in a new form, elevated to a higher level. In this sense, the "spirit" comes back to itself, picks itself up from pieces to *re*construct a rational whole: "It wins its truth only when, in utter dismemberment, it finds itself."[13]

Perhaps this sounds a bit mystical as it stands, so let us look at an example. When Greek mathematicians discovered that the square root of 2 cannot be expressed as a ratio of integers—which was not a small scandal—mathematical knowledge was not pulverized. The problem was resolved by Eudoxus, who introduced a new methodology that utilized a geometric approach, in essence employing the idea of approximation, "the method of exhaustion." Pythagoras's number mysticism and the arithmetical bias of Pythagorean mathematics eventually gave way to a different, geometric, dynamical viewpoint. What was at first a logical scandal gave way to a reform that accounted for new mathematical "realities." The spirit of mathematics won its truth: It found itself in despite having been annihilated, negated, by the discovery of irrational numbers.

This is in a certain sense an exemplary case of the process of negation according to Hegel. For Hegel, negation is a basic operation, but not in some static sense. It is a process of negat*ing*, with various things, such as time and memory, inscribed into it. An important feature of this endless dynamism of negation is what Hegel calls "sublation" (*Aufhebung*). This crucial Hegelian coinage evokes not only annihilation but also overcoming, preserving something on a higher level. Every concept invokes its own negation, and this opposition is then sublated into a higher unity. This "triangular" structure of thesis-antithesis-synthesis then recurs on a higher level, and then again on ever higher levels, like new triangular structures in the construction of Koch's snowflake curve, in a dialectical process of concept formation.

This dialectical process is extended in history, but Hegel believed that by looking at it from a certain vantage point, it could be regarded as complete. We are certainly not talking about an ordinary kind of completeness in this case. For Hegel, knowledge is not the knowledge *of* something fixed, as in the classical logic of timeless, immutable objects, but rather a process of reflection that by its very nature involves questioning, negation, annihilation, and elevation to ever higher theoretical ground. Absolute knowledge is the knowledge of knowledge itself, not as some ultimate *thing*, but as an understanding of *how* it works and changes in time.

We have here a new form of the old Heraclitean riddle: Everything is in flux, everything changes, but what always persists through all changes is the change *itself*. For Hegel, knowledge persists through changes precisely because it understands this "paradox." History does not bother *it*, because it understands the annihilating power of time as the very condition of its movement toward its triumph, which consists in always reconstructing itself anew. The fall is necessary for the knowledge gained in the recovery. It is in this sense that the spirit can be "complete" for itself (though not for us). It understands the necessity of its own transformations and thus becomes part of the Heraclitean change that persists through all changes.

Hegel's position has been the subject both of overmuch praise and of truly severe dismissals. But we will not worry about that. For the time being, my only concern is to introduce some terminology and mark a few philosophical reference points. Now that we have that at our disposal, let me bring a little mathematics into the picture.

3
SPACE ODDITY AND LINGUISTIC TURN

It would seem that the conflict between science and continental philosophy can be traced at least as far back as Kant. With varying intensity and in very different forms, it then extends through the unbridled romanticism of the nineteenth century and all the way into the present.

Mathematics, specifically geometry, has had its hand in this conflict. I would therefore like to place the issue between Kant and geometers under the magnifying glass and look into the possibility that it had a role in the development of an "attitude" toward continental philosophers' understanding of science and mathematics.

Here are Kant's famous words on judgments in geometry: "That the straight-line segment joining two points is the shortest path from one to the other, is a synthetic proposition."[1] In other words, our knowledge that the shortest path between any two points is the straight-line segment that joins them is based on the a priori intuition of space. We are given to conceive of space in this way and, at least on some readings of Kant, in no other way.

The statement has fairly specific mathematical consequences. The space in which the shortest path between any two points is the straight-line segment that joins them is in a sense "flat," without any curvature. It is known as Euclidean space. This is the space whose geometry we studied in high school, represented in the geometry of ancient Greeks, where angles of triangles always add up to 180 degrees. And it *seems* to be the geometry that we ordinarily experience visually. So Kant now appears to be claiming that the a priori intuition of space forces upon us the notion that the space, as far as we can tell, is Euclidean. In particular, it seems that he is trying to say that Euclidean geometry is the only geometry knowable to human beings.

Let us glance at Euclid's *Elements*. Euclid starts with definitions, axioms, and postulates and thus builds his theory. But there seems to have been some dissatisfaction regarding the fifth postulate—precisely the one that singles out what we call Euclidean geometry:

P1. Each point can be joined with any other point by a straight line.
P2. Each line segment can be continuously extended into a straight line.
P3. From each point it is possible to describe a circle with any radius.
P4. All right angles are mutually congruent.
P5. If a straight line falling on two straight lines makes the interior angles on the same side less than two right angles, the two straight lines, if produced indefinitely, meet on that side on which the angles are less than the two right angles. (I.e., whenever there is a line L and a point P not on that line, then there is exactly one line parallel to L that passes through P.)

Even on the purely lexical level, the discrepancy between the fifth postulate and the others seems quite clear. The fifth postulate is certainly as complicated as some of the assertions to which Euclid gave the status of theorems. The fact that it thus stands out (although it is possible to formulate it in a simpler way) fueled the imagination of many a geometer: Perhaps it *is* a theorem. Its fatal appeal was legendary well before Kant. Many geometers have addressed the issue over the centuries, but one of the most dramatic cases seems to be that of the Italian geometer Girolamo Saccheri. He worked very hard to prove the postulate, published a flawed argument in his *Euclid Cleared of Every Flaw* (1733), and then died the year it was published.

About a hundred years later, the fifth postulate was found to be logically independent of the remaining ones: It cannot be deduced from them. One way to show this is to provide a *model* for a geometry where the fifth postulate fails while the rest of the assumptions hold true. To provide a model means to assemble a collection of geometrical objects that we could call "points," "lines," "circles," and so on, where all these objects satisfy all of Euclid's axioms and postulates except the fifth. In other words, we find an alternative interpretation of geometrical concepts that would conform to all the postulates except the fifth. This was done within the "Euclidean" space we ordinarily seem to be experiencing—apparently contradicting Kant's claims. Thus, Kant became vulnerable to a strictly mathematical critique.

German mathematician Karl Friedrich Gauss might have been the first to voice his dissent with Kant, at least privately, in an 1817 letter to a colleague. Around 1826, Russian geometer Nikolai Lobachevsky constructed a model of a non-Euclidean plane. Further confirmation came in 1832, when the work of the Hungarian engineer and geometer Janos Bolyai was published. By 1840, Lobachevsky's work appeared in German, and the news was slowly being accepted.

Gauss was aware of the implications of such research. In an 1832 letter to Bolyai's father, Gauss wrote: "It is precisely in the impossibility of deciding a priori between Euclidean and [young Bolyai's non-Euclidean geometry] that we have the clearest proof that Kant was wrong to claim that space is only the form of our intuition."[2] And so, apparently, goes Kant's philosophy of geometry.

A well-disposed observer could grant Kant some latitude and agree that he simply misjudged the details of what it is that intuition gives us about space. One might even say that he followed his own advice and "submitted his observations to a jury of fallible men." Finally, the possibility that he was interested in showing that we *at least* have the knowledge of Euclidean geometry—rather than demonstrating that it is the *only* geometry knowable to us—does not seem altogether inconceivable.

But that is of little consequence now. Kant was thought to have made an error, and sciences proceeded to "correct" not only his mathematics but his entire philosophical outlook. The sequence began with experiments conducted by German physicist and physiologist Hermann von Helmholz. Helmholz held that spatial orientation is acquired and that Kant confused the acquired ability for "unconscious inferences" with an innately given intuition of space. In his 1870 article "On the origins and meaning of geometrical axioms," Helmholz announced that it is possible to have a visual experience of non-Euclidean space.

In 1913, there was an even more "damaging" experiment. Subjects were asked to arrange objects ("points"), first into two lines that are parallel, then into two lines where corresponding pairs of points would be equidistant. In Euclidean geometry the concepts of parallel and equidistant lines coincide. (Rails on a railway are parallel, and one reasonably expects that the distance between won't shrink as one travels along them.) So the resulting arrangements should have been the same. But they were different. This indicated that "visual space" of the tested subjects may not be Euclidean. Since 1947—the time of other, more extensive tests—it seems to be accepted that the geometry of our visual space is not a "flat," Euclidean one, but has varying curvature.[3]

I certainly do not want to make too much of this collision between Kant and the sciences, because a good deal of it consisted of earnest attempts to clarify the problems that Kant opened up. Helmholz is sometimes described as being "neo-Kantian" in some sense. Einstein paid tribute to Kant, although he argued, more on the romanticist line of thought, that the imaginative-intuitive constructs that connect concepts and experiences cannot be articulated within an unalterable scheme.

However, it is difficult to ignore this "clash" in light of the fact that some indubitably excellent historians of mathematics cannot resist employing a delicate form of irony when it comes to the topic of Kant and geometry. For example: "[D]espite never having been more than forty miles from his home city of Königsberg in East Prussia, [Kant] presumed he could decide the geometry of the world."[4] If a threshold quantity of air-miles is required to enter the debate, one naturally wonders what should be said of Galileo and Nicholas of Cusa—they are credited with introducing the idea that space is *infinite*. It is of course true that they did not claim to have established this by introspection, but it is difficult to imagine that they had incontrovertible empirical evidence for their hypothesis.

This is not a trivial problem. For example, American mathematician Jeffrey Weeks has recently published several articles in which he examines ways of testing the hypothesis that the universe is finite but "with no boundary" in a certain technical sense. (Interestingly, already in Descartes one can find a pedantic distinction between space as "*un*bounded" and space as "*boundless*.") Weeks published *The Shape of Space* in 1985, produced computer-generated videos that illustrate the possibility that the universe is not infinite even though it may look infinite to us, and received a prestigious prize for his work. It might not be "mainstream" science—whatever that means—but the issue is certainly alive.

Perhaps we should consider the case of the boundedness of space in more detail. Apart from being a problem of more than antiquarian interest to contemporary cosmologists, it will serve to illustrate some important internal conflicts within Kant's philosophy, ambiguities that at least to some extent reflect the intellectual climate of the time.

Let us leave aside the data obtained from space probes, which Weeks intends to use to determine the geometry of the universe. Instead consider a more modest question: How would I, an ordinary guy who has no connections at NASA, *know* that space is infinite? It is true that I can imagine myself as moving away from Königsberg in East Prussia toward Fredericton in East Canada, and further on to even stranger places. That proves little—if anything, that it is not impossible for space to be infinite. But I could also argue, in the spirit of the old Pythagorean doctrine, that is not impossible for the world to be a sphere. For instance, I could imagine that as I move toward the "boundary" of the world, from some place inside the sphere toward its surface, everything inside the sphere is getting smaller in equal proportion: me, my measuring stick, and all else. I would not notice a thing. The results of my measurements would be just the same *as if* the world were infinite.

I seem to have a situation now. Pure reason, as in the case of the somewhat trivial example with round squares, lets me derive two mutually incompatible conclusions. On the one hand, space seems infinite; on the other hand, it could well be finite. I cannot know which of these conclusions is correct unless I view the totality of space from outside of space itself, which I cannot do.

Kant's resolution of this antinomy is in a sense similar to the resolution of the example with round squares. Reason permits me to draw incompatible conclusions because I am not reasoning about a phenomenon, an object of my consciousness. Space, as a totality, is *not* an object of my judgments, at least not of reliable judgments. It could exist as a thing in itself; Kant does not deny this at all. But as far I as am given to *know*, this totality is, so to speak, the work of my imagination. The totality of space (as an "infinite given magnitude") is a necessary illusion, a figment of fantasy that is nevertheless a *precondition* of knowledge. Indeed, Kant describes imagination in general as an "indispensable function of the soul, without which no knowledge would be possible."[5] Such ideas as that of the totality of space, Kant says, are a natural and inevitable illusion.

One could even say that the appearance of non-Euclidean geometry vindicates Kant's ideas to some extent, despite having "exposed" him as less of a geometer than was Gauss, Lobachevsky, or Bolyai. Space, as a *thing-in-itself*, could be Euclidean, or it could be non-Euclidean. As a *thing-in-itself*, it could be finite or it could be infinite. Science overcame this dilemma in a "Hegelian" way: Galileo annihilated the Pythagorean notion of the spherical universe and replaced it by the abstract infinity of Euclidean geometry, but the opposition between the two views was later dialectically "sublated" into a theory about the universe that is expanding, bubbling, and so on.

But from the Kantian viewpoint, which takes the individual human being as the central measure of things, I am not in a position to know whether the "totality of space" *is* this way or the other. As an ordinary human being who experiences things locally—*not* as a scientist equipped with all sorts of powerful instruments and fancy theories—I cannot make a knowledge claim about a pseudo-object that is not within the realm of objects of my experiences.

It is sometimes neglected in "scientific" accounts of Kant that he was brought up in a pietist atmosphere, and that his philosophical outlook reflects a form of anti-elitism. In the *Critique of Pure Reason*, he wrote: "Do you really require that a mode of knowledge which concerns all men should transcend common understanding, and should only be revealed to you by philosophers?"[6] Kant was concerned with a theory of knowledge that would be equally accessible to everyone, rather than custom-made to fit scientists, philosophers, and other academics.

The problem of totalities is not something that pertains only to physics and geometry of space. It is more general than that. Consider the so-called "paradox of the ravens." Suppose that I am trying to confirm the hypothesis that all ravens are black. It seems relatively natural to say that the likelihood that the hypothesis is true increases each time I run into a new raven that happens to be black. But my hypothesis is logically equivalent to the assertion that all nonblack things are not ravens. Hence, I can lend additional "confirmation" to my Theory of Ravens by performing an experiment that shows that I have several white shirts at home that are not ravens. I seem to have founded an entirely new scientific discipline: indoor ornithology.

That seems farfetched, but my method is logical. The hypotheses "all ravens are black" and "all nonblack things are not ravens" are logically equivalent. Yet this equivalence leads to something that appears absurd. There are certainly ways of resolving this paradox. They range from denials of the possibility of confirmation, to sophisticated theories about experimental confirmation. But since there are disagreements about various resolutions of the raven problem, and since we are talking about Kant, imagine what might be the "Kantian" view of the antinomy of ravens.

The problem, or at least part of the problem, is that I am imagining the world as an object, a totality I could somehow freeze-frame, in which I could round up all the ravens and determine that they are black. If I could freeze-frame the world and examine everything in it, then it would not matter whether I am proving that ravens are black or that nonblack things are not ravens. But I cannot do that, because I am *in* the world.

The point, roughly speaking, is that the idea of an objective totality seems to imply a subject for whom this totality would be meaningful. The meaning I attribute to a "totality" involves me in some way, which is to say that it might not be as objective as I had hoped. This is not to say that the entire universe is my "psychological construct" that vanishes when I go to sleep. What Kant is trying to say is that, from the point of view of the individual human being, the universe is not the same *kind* of physical object as a book, although I imagine it *as if* it were.

This appears to be the fundamental scandal of Kant's antinomies. Reason can be deceptive when it exceeds its legitimate authority and applies to "things" that are not the objects of our experiences—empty ideas, imaginative constructs, or things-in-themselves—and yet it cannot help doing just that. Kant, in the *Critique of Pure Reason*, wrote:

> The perplexity into which [reason] falls is not due to any fault if its own. [It] begins with principles which it has no option save to employ in the course of experience, and which this experience at the same time abundantly justifies it in using. Rising with their aid [. . .] to ever higher, ever more remote conditions, it [. . .] finds itself *compelled* to resort to principles which overstep all possible empirical employment [. . .]. But by this procedure human reason precipitates itself into darkness and contradictions.[7]

Kant could just as well have said thus about the so-called "new math," but let us stay on topic. Reason is compelled to produce certain "transcendental illusions." Given Kant's relatively precise use of the word "transcendental," this would seem to imply that such illusions are the necessary condition of the possibility of knowing. Our acts of synthesis produce unifying fictions, "ghosts" that are never given (*gegeben*) but are necessarily posited (*aufgegeben*).

Let me take advantage of this opportunity to introduce an idea that I believe is an important ingredient of the cultural clash we now call science wars. I think that part of it stems from a difference in esthetic sensibilities. It has been argued by various people that the esthetic sensibility of modernism consists precisely in the shattering of illusions, those "ghosts" of which I spoke in the preceding paragraph. This goes for art as well as for science, because modernist art shares the esthetics of modern science. I do not mean only the canvases with squares and circles on them, but something more general than that: the removal of "esthetic fictions" and their replacement by a more rational approach, in a sense more akin to science and mathematics. The canonical example of such a "modern" disposal of illusions occurred in the sixteenth century, when Copernicus dismantled the Geocentric Dogma. It is only through an effort of reason that this could be done, because most of us *see* the sun rising up and down *as if* it were circling the earth.

Science, in this sense, stands for the ideal of the modern era. It is an effort aimed at undoing various illusions, fictions, and deceiving intuitive "insights" of all kinds, of which Kant's intuition of space is only one item on a much longer list. Modernist art raised this notion to an esthetical attitude, but this variety of esthetics goes back to Galileo. (He is known to have said that reason can "rape the senses," and it is probably for the same reason that modern art can on occasion seem a little challenging to the eye.)

This esthetic sensibility is expressed nicely in Galileo's commentary on the work of the sixteenth-century Italian writer Torquato Tasso. In *Considerationi al Tasso*, Galileo—an art critic of considerable repute in his day—judged Tasso's penchant for allegory as suffocating. It suffocates, wrote Galileo, with mannerisms, with chimerical and completely unnecessary fantasies.[8]

Thus, we seem to be uncovering a potential source of cultural conflict. Kant pointed out the necessity of some illusions in the thinking practices of ordinary human beings. Science and the culture of modernism insist on doing away with them. Romanticism followed Kant's emphasis on the importance of imagination, radicalized Kant's notion that the "esthetic idea" is something ultimately inexhaustible by reasoning, and emphasized the importance of metaphors, fictions, and allegories—precisely the things that Galileo found suffocating in Tasso. Romanticism and its descendants sometimes took things to extremes that Kant would probably have regarded as *Schwärmerei*: dreaminess, exaggeration, confusion. Science and the culture of modernism went to the opposite extreme and dreamed of the ultimate positivist explanation of everything.

Various attempts to remove all esthetic fictions and imaginative constructs, attempts that in some sense characterize the "modern era," were not altogether successful, and it is not clear that they ultimately can be. One does not have to look toward continental anti-rationalists to find that the complete removal of all fictions can lead to no end of trouble. Even a utilitarian like Jeremy Bentham—Kant's British contemporary, approximately—came to what could be construed as an analogous conclusion.

Bentham attempted to analyze away fictitious entities from language and to treat sentences as logical constructs upon actual experience. In the process he came to distinguish imaginary objects as either nonsensical nonentities (winged horses, unicorns) or symbolic constructs (one speaks of a "social contract," although no contract has been signed). Some complications arose along the way and Bentham, not entirely to his delight, found that it may well be impossible to speak about reality in a sensible manner without invoking a certain irreducible type of fiction.

Fictions, for Bentham, are "those sorts of objects which in every language must, for the purpose of the discourse, be spoken of as existing." And he added: "I know that fictions are unreal, but I nonetheless speak of them as if they were real objects."[9]

It is to language, says Bentham, that fictitious entities owe their "impossible yet indispensable" existence. They are, on the one hand, the source of confusion and should ideally be banished, but are on the other hand an indispensable tool. The "objective" totality of the legal system is such a fiction, and one generally avoids visiting places where this illusion is shattered.[10]

Already in the days of Kant and Bentham, we can sense the outline of possible strategies for dealing with the difficulty of these necessary fictions. Bentham may well have been the first to attempt to remove these entities from language by classifying them, initially, as fictions of the first degree, second degree, and so on. Bentham's strategy is based on the idea that all meaningful sentences of a language are reducible to logical constructs upon immediate experiences (which, I suppose, are fictions of degree zero).

A revival of Bentham's approach became known in the twentieth century as logical empiricism or logical positivism. In addition to other difficulties of this doctrine, it seems to leave mathematics in a problematic position. The fictional constructs of mathematics—points, space, the infinity of arithmetic, to mention only the milder ones—stubbornly resist being brought down to experience.

A possible way to deal with this would be to fall back on some earlier attempts to treat mathematical activity as pure symbolic manipulation, to view mathematics as a language whose meaning is not in its relationship with reality but is somehow self-contained and discovered by suitable methods. This appears to be Leibniz's idea: "[F]rom the sole structure of the expression we can reach the knowledge of the properties of the thing expressed."[11]

The influence of this idea is almost impossible to underestimate. It seems that various twentieth-century attempts to treat language as an algebra of symbols—which derive their meaning not from what they are but solely from their structural relationships—owe much to Leibniz's notion of "blind thought" (*cogitatio caeca*, a kind of symbolic manipulation). In its extreme form, this approach leads to the doctrine that meaning springs from symbols either by itself or as the result of applying a vaguely mechanical, automatic method.

But while it is true that Leibniz dreamed of making "language and calculation the same thing" by formal operations on "characteristic numbers," it is also true that he acknowledged that he was dealing with fictions: "I pretend that these marvelous characteristic numbers are given [. . .]."[12] Leibniz, the philosopher of preestablished harmony, pretends.

Let me now turn, if only as a somewhat dramatic counterpoint to these eminently "modern" programs, to some ideas of Kant's German contemporary Johann Gottfried Herder. For Herder, as it was perhaps for Vico and even to some extent for Kant, "mind is shaped by fictions."[13] Rather than attempting to subdue or remove such fantasies, rather than subscribing to the pragmatic attitude of proper measure or of Kantian a priori categorization of our formative fictions, Herder seems to have believed that it is desirable and indeed necessary always to express oneself, to use one's imagination to the fullest, and thus to unleash all the creative powers of the human mind even at the cost of having to sacrifice the notion of ultimate and objective truth. It is only freedom that counts, especially the freedom to make the truth.

But if the truth is the made (as Vico says), if truth is in some manner a linguistic expression of our "grounding" fantasies, then it is not at all clear why such formative fictions—even if they *were* mutually compatible—should always have compatible linguistic expressions, or be forever fixed. Thus, it may be difficult to *justify* their universality. At best, they may appear to be common to a linguistic community, to a community of people who, to put it in the more contemporary terms that are usually ascribed to the later work of Wittgenstein, share a particular "form of life," that is, share the same fantasies and necessary illusions, expressed in their linguistic practices.

One cannot easily dismiss this notion, even in the case of mathematics. For instance, a number of years ago, an important theorem was proved in a joint work done mostly by Anglo-American mathematicians. So many people were involved in proving it that it was not clear whether any one person actually understood the whole thing. For this reason, it was regarded with some suspicion in Russia, and the editors of the annual report on problems and progress in algebra (*The Kourovka Notebook*) felt compelled to mark the problems solved using this theorem by a special symbol.

That is not to say that they did not believe that the theorem was indeed *true*. As the Russian mathematician Sergey Adian once said, "I believe, but I like to check." The editors did not mark the problems whose solution invoked this theorem as *un*solved; they only put a discreet asterisk next to them to indicate that the solutions are in some sense special. Their "form of life," their mathematical practices and mathematical culture, caused them to be cautious with regard to the form of the theorem's linguistic *justification*, that is, the proof. (This is not something peculiar to Russia. There are similar examples elsewhere and there is an interesting problem regarding the acceptability of computerized proofs. But I hope the point is made. It does not follow that mathematics is "determined" by culture. However, culture, practices, and politics do in some way enter the picture.)

Let us return to Herder. From this initial point at the turn of the nineteenth century, there extends—it seems, all the way into "postmodernism"—a very different notion of truth. Truth cannot be separated from language, community, and history. Thus, the truth of others, say, of ancient Greeks, can be understood only partially and incompletely, insofar as we could place ourselves in their historical and linguistic position. All truth involves an act of interpretation, which might involve individual imaginative contributions, beliefs and practices of a community, and other nonobjective things. This instantly brings into question the legitimacy of applying our "universal" standards to other cultures. Even the truth of my own community remains unstable, and I must always express myself to reconfirm my *belonging*—a crucial term for Herder—to something larger than myself. I am always in need of the "other." All of my meaning in essence arises through communication and interpretation, through a lived experience that cannot be reduced to an a priori fixed "scientific" category. It is in language that I must realize my freedom and my creative capabilities. (This is why Herder complained that Kant neglected the problem of language.)

Without the language and the community, without this "other," we are, to paraphrase another romantic thinker, signs without significance. Therefore, my freedom is in my inexhaustible capacity to reinterpret, imaginatively recreate, that which comes to me by way of language. But if all of us do that, we lose the certainty of communication. Unlike nineteenth-century science, romanticism was willing to consider that price.

Most of these ideas can be traced in one form or another right back to Königsberg in East Prussia, though not to Kant but rather to a man whom he knew, respected, and probably also regarded as a confused mystic: Johann Georg Hamann, *der Magus in Norden* (the magus of the North). Hamann held that the true reality, the whole of the actual lived experience, cannot be categorized, subdued to reason, analyzed, and dissected: "What is this highly praised reason," says Hamann, "with its universality, infallibility, overweeningness, certainty, self-evidence? It is a stuffed dummy which the howling superstition of unreason has endowed with divine attributes."[14]

It is only through an "unreasonable," imaginative individual act of synthesizing experience into a seamless symbolic whole that reason may get a chance to offer its analyses. For Hamann, and for a good deal of nineteenth-century romanticism, restraining this imaginative power of the individual, analyzing it in "scientific" terms with the intention of giving it an unalterable description, becomes the equivalent of a maiming, a depriving of a fundamental freedom—even a spiritual death of sorts.

Against the background in which these notions were stewing, it would seem that Kant was concerned with reconciling the universality of truth with the central place of the synthesizing power of individual consciousness. Kant's philosophy could then be thought of as a critique of the analytic excesses of rationalism (he wrote of the "ridiculous despotism of the Schools") as much as a critique of Herder's and Hamann's notions (which he seems to have regarded as ungrounded speculation, with all due respect). If this is the case, Kant may well be seen as having attempted to prevent a "science war" all the way back in the late 1700s.

There are also those who argue that Kant's intervention was not intended to resolve or reconcile anything, but on the contrary to point out the *necessity* of bringing such conflicts into the open. In a different but not entirely unrelated context, in a piece on the idea of the university—with the eternally relevant title *The Conflict of Faculties*—Kant wrote: "This conflict can never end, and it is the philosophy faculty that must always be prepared to keep it going."[15] If that was Kant's objective, then his program, although scarred by its "space oddity" or even *because* of it, may well be considered a success.

Kant's Euclidean adventure in the geometric wonderland appears in the end to have fueled the disagreement between romanticism and science. "In a sense the discovery of non-Euclidean geometry," wrote Carl Boyer in *A History of Mathematics* (1989), "dealt a devastating blow to Kantian philosophy [. . .]."[16] In a sense. But perhaps we should not get carried away and blow our own importance out of proportion. There were quite serious critiques of Kant before 1826—by various romanticists and by Hegel—although it is clear that mathematics played a role. Mathematics, we could say, provided a concrete counterexample where others had already pointed out a gap in the proof.

In any event, exercising some caution regarding the term "Kantian philosophy" is probably not a bad idea. For instance, according to the widespread cliché, various post-Kantian romanticist thinkers are supposed to have attempted to ground all knowledge in the knowledge of the self, which would allegedly be accessible to them by means of "intellectual intuition," a mind's inward-looking eye of some kind. Well, some of them thought something like that, some didn't, and some vacillated.[17]

Nevertheless, science, and various branches of philosophy that align themselves with science, dismiss romanticists out of hand as a collective, due to the deceptiveness of introspection. Kant's mathematical blunder is considered paradigmatic of such deceptions and as such is used as an argument against everything that comes close to Kant, which includes romanticism. It seems that it is from this somewhat swift generalization—in addition to the more basic ideological disagreements over romantic "holism" versus scientific "reductionism," and the clash of esthetic sensibilities—that at least a part of the misunderstanding between science and continental philosophy stems.

It may well be the case that understanding the romanticist point of view would not in the least change the mind of the scientist. Still, it would be nice to be able to make an informed decision. It does not seem to be the case that romanticism—not all of it, at any rate—sought the certainty of knowledge within the self.

Fichte, to name only one in the long line of philosophers who can in broad terms be associated with this tradition, criticized Kant precisely for attempting to give the prelinguistic familiarity that consciousness has with itself an a priori determined "scientific" description. On the romanticist view, I cannot know myself, at least not in any standard sense of the word "know." I can only *conceive of* my "I" as a kind of continuous creative action that cannot be captured in an a priori fixed language. Thus, even if I had some "knowledge" of myself, that "knowledge" would be beyond a final linguistic formulation and as such could not be used to ground a priori justification of anything—least of all a specific kind of geometry.

With that in mind, one could even say that the discovery of non-Euclidean geometry supported rather than devastated the philosophy of the romantic tradition, insofar as these thinkers cared about geometry at all. At any rate, many romanticist ideas survived throughout the 1800s and influenced several mathematicians, philosophers, and writers, at the turn of the twentieth century. It is in the mathematical expression of some decidedly romanticist notions—the inability of logic and language to capture the sense of continuity, in particular the continuity of inner time, as well as the doctrine that no language can be said to be the guarantor of certainty of its own meaning—that we should look for a part of the connection between mathematics and postmodern thought.

Before considering relevant mathematical arguments, let me try to sum up what I have "established" so far. Despite the clash of esthetic sensibilities between science and the culture of early romanticism, it appears that romanticist thinkers managed to place two important issues relatively high on science's agenda: first, *language*; second, the problem of *continuity*—that ineffable inner flux, the sense of continuous creative action—and its relationship with language.

Kant's geometric controversy seems to have contributed to the urgency of reconsidering the notion of mathematical rigor, of formalizing and arithmetizing all of mathematics, and of questioning the idea of self-evident truths. In *Mathematics—The Music of Reason* (1992), French mathematician Jean Dieudonné wrote: "If I wished to sum up in one sentence the way in which ideas unfolded during this period [1800–1930], I would say that its essence was a progressive abandonment of the concept of 'evident truths,' first in geometry, and then in the rest of mathematics."[18]

Mathematics was making its own "linguistic turn." With the inertia of a luxurious ocean liner, it was veering toward the discrete, formal-computational approach that reflected a growing concern with language.

4
WOUND OF LANGUAGE

To free the mind from the despotism of the eye.
—Samuel Taylor Coleridge

Before we look into the objections that mathematicians addressed to the reduction of mathematics to logic, it might be instructive to devote a few pages to the elaboration of some romanticist doctrines. (I will not insist on a rigorous distinction between the culture of romanticism and the philosophy of romantic idealism. The affinity between the two allows for taking such a liberty.)

The mathematicians whose arguments I then outline invariably invoked metaphors that compared mathematical understanding to an artistic, creative process irreducible to the "sterile" methodology of pure logic. They objected to the notion that the total meaning of a (mathematical) text is somehow encoded in the text itself and is subsequently recovered by a quasi-mechanical process of "decoding." In this sense, they can be placed within a broadly conceived tradition of romanticism, one of whose tenets is that *every* act of understanding has something of an artistic quality, a certain uniqueness and individuality.

As a convenient starting point, consider Coleridge's locution "to free the mind from the despotism of the eye." The metaphor can be read in many ways, but Coleridge may have intended the liberation of the mind to mean something like this: What gives some object an artistic quality is precisely its ability to evoke a transgression of the observed. One divines the intention of its creator, the deep well from which it came into being. In a similar vein, Novalis wrote that a poem must be inexhaustible—as inexhaustible as is man himself. There is more to art, infinitely more, so to speak, than meets the eye. It is an invitation to a dialog, an opening toward the *un*seen, unforeseen, and unforeseeable.

This invisible "remainder"—or rather the gap between the finite physicality of artwork and the infinity of possible interpretations, the gap between looking and seeing-as—opens up the playing field to irreducible individual experiences and self-realizations of the observer (reader). It is the interpretive process that associates art with life and vitality and liberates the mind from the despotism of the eye.

That seems to be how romanticism reinterpreted the ideal of classical esthetics, the ideal of "art becoming life." The unending search for the essentially unformulable meaning of art lets me realize myself through a kind of creative evolution by employing my sense of beauty, imagination, or at any rate some extra-logical faculty of the self that can be felt but not fully known.

It might be too much, even for proverbially absent-minded mathematicians, to *identify* mathematics and art. Nevertheless, the underlying strategy of mathematical critiques of logical reductionism has been to elaborate the analogy to a certain extent. On this view—in sharp opposition to the views of various logicians—mathematics is something beyond mathematical *text* or a "mechanical" decoding of a text. It is a human activity of which the text is at best an imperfect record, a guide for the creation of mathematical meaning.

An important figure of early German romanticism, Friedrich Schleiermacher, formulated this idea with particular clarity in the context of art. Artwork is something abstract that is not anything *in itself*, but rather becomes something only through the relationship with the observer. The observer is thus granted a certain degree of interpretive autonomy. The meaning of artwork is not simply discovered or "decoded." It is always created anew. (There is a tendency to wear out the vague term "creativity" in a discussion of romanticism. I will keep using the term, but keep in mind that already Descartes scoffed at the notion that truth can spring from one's head "like Pallas from the head of Zeus.")

By means of such an interpretive process the viewer's mind is confronted with the artist's mind, so art is in a sense a form of communication. In addition, this interpretive process enables one to learn about oneself in relation to "the other." The play of interpretation is thus a way of finding out about oneself and others, a way of expanding one's imaginative horizon. This reflexive scheme is known as the hermeneutical circle. The hermeneutical circle is never complete: It is an unfinished process, a helix, as it were, of interpretation.

Something similar is true of the romantic understanding of self-consciousness. My ego, "me," the knowing-I, can come closer to the unknowable "I" only by means of a search through the inexhaustible world of possibilities for outward expression of my inner self, the will, the play-impulse (*Spieltrieb*), or at any rate some creative power, a spirit at work within me. One spoke of it as the "philosophy of finitude." It viewed the I as a relation of self-reflection grounded in something that is itself *not* reflected, something that cannot be known.

For example, German philosopher Friedrich Schelling criticized the Cartesian tenet "I think, therefore I am" for just this reason. One might as well say—he notes of Descartes—"I digest, I produce fluids, I walk, I ride; therefore I am."[1] The ego does not speak for everything that I am, although it likes to claim that it does. And so, says Schelling, the ego-logical *me* is simply a mode of *appearance* of the I.

There is something that precedes my (intellectual) reflection upon myself, and I am "absolutely dependent" on it. It can be sensed only as something that "resists" the larger, more inert context of the material world, culture, or language, none of which are my own doing but which precisely because of that I must endlessly reinterpret, make my own, enrich with my imaginative powers. Then I can *feel* that I am a living, free, continuously moving force that exceeds all linguistic edifices (which are relatively inert).

Thus romanticism is fundamentally concerned with the inexpressible. It typically relies on symbolism and allegory to evoke a certain unembracable whole that is its euphoria and its agony. "Your relation to the universe is inexpressible," says Isaiah Berlin of romanticism, "but you must nevertheless express it."[2] To be an artist is to live and to interpret, always to act in order to creatively realize yourself, to push onward with your metaphorical descriptions even if you believe that they are necessarily incomplete because they merely impose a stagnant form on something that is formless or at least always in flux.

A fairly unrestrained variant of this viewpoint was expressed in the remark attributed to the German philosopher Friedrich von Schlegel: "No, it can never be seized because mere imposition of form deforms it."[3] Along with other romanticist ideas, this would later be embraced and pushed to extremes by Friedrich Nietzsche—a dropout, so to speak, from the school of romantic humanism—who maintained that the world is "a monster of energy," "a becoming that knows no satiety," a "*Dionysian* world of the eternally self-creating,"[4] which can never be captured into the net of logic.

Similar ideas, much closer to romanticism than is Nietzsche's vision, would also appear in the work of the early-twentieth-century French philosopher Henri Bergson. Bergson held that pure reason artificially immobilizes everything and thus always fails to describe the incessantly changing flux of lived experience. Immutable truths and definite form, much liked by science, are for romanticism and its descendants something akin to death. The only relief, however temporary, however ephemeral, that romanticism allows itself in its struggle with the fleeting flux is the relative inertness of a myth.

Reason, in particular, is itself a myth. That does not mean that everything is arbitrary and random. But there are things that go beyond reason, things that cannot be fully expressed linguistically or logically. This especially goes for that continuous inner flux, an activity that always eludes complete description. In a certain unbridled form of romantic thought, falling prey to the lazy belief that reason can actually impose a definite form on anything is to be a sinner of the highest order, to be "dead," as it were. "Art is the Tree of Life," wrote the mystical poet William Blake; "Science is the Tree of Death." It might be interesting to record another quip of Blake's at this point: "God forbid that Truth should be confined to Mathematical Demonstration!"[5] In a sense, the sentiment would be shared by a number of mathematicians who objected to the reduction of mathematical truth to formal demonstrability.

Against those who disagree, romanticism has a powerful weapon: irony. To counter every conceivable form of ossification, any unimaginative and rigid or freeze-frame view of the world, a romantic employs mockery, blows it up by playful reinterpretation. Yet this powerful weapon is at the same time the source of what from the rationalist point of view seems like a profound flimsiness that simply cannot be remedied. If every act of understanding is in the end the artistic and creative work of the logically irreducible individual, then the problem arises of how any universality, in particular that of science but also of meaning, can be justified.

One cannot complain that romantic thinkers were not aware of the sensitive nature of this issue. Schlegel, for instance, wrote as early as 1800 that critical reflection on understanding reveals an amount of "positive nonunderstanding" and that all reflections on language must involve an ironic awareness of the "impossibility of complete communication."[6] Wilhelm von Humboldt, a language scholar who together with Schleiermacher was one of the founders of the University of Berlin, said something to the effect that no one means the same thing as anyone else, and that the slightest difference trembles in language like a ripple on the surface of water.[7]

The early romantic idealists developed sophisticated theories for resolving this difficulty. Their proposals, however, did not seem to exert much influence on the culture of late romanticism. Instead, these theories gave way to two sentiments that I address because of their connection to the popular culture of "postmodernism."

The first of these sentiments is nostalgia, a longing for a lost "golden age" or the age of the noble savage, when things were simple or at any rate simpler, harmonious, guaranteed by God and by geometers—or by the god-geometer—to remain that way. Things are different now. Odysseus cannot come home anymore except in the sense that the endless journey itself, rather than Ithaca, is his home. "Philosophy," wrote Novalis, "is in essence homesickness. A longing everywhere to be home."[8]

The second sentiment is that of paranoia, a sense of always being intruded upon by language, of being misconstrued by others, of being tricked by the "cunning spirit" (a notion sometimes attributed to Hegel) whose omnipresence one can never escape. Thus, I am always implicated in things that are not my own, which furthermore have the power to turn my utterances and my intentions into something altogether alien to me. Every sentence I construct—as Jean-Paul Sartre would write much later—seeps away from me; my words are being twisted while they are still in my mouth.

This second sentiment, this "wound of language," is of particular interest to us. I am of necessity thrown into language, or another form of symbolic order, in which others threaten to take my meaning and turn it around, toss it back to me changed and transformed. I must act, continually express myself, but every "passage to the act" could end up having unintended consequences—like the proverbial Chinese butterfly whose innocent wing-fluttering causes a nasty blizzard in Canada.

This can lead to a kind of mystical vitalism, the idea that it is the language-spirit or nation-spirit that speaks *me*, rather than the other way round. By extension, I am not bird-watching but am being watched by birds. The world acquires a disturbing atmosphere of an Alfred Hitchcock movie in which I am being gazed upon by the eye of some entity from whose despotism my mind can never be freed.

Let us leave these slightly paranoid tendencies aside for now and formulate the problem in the following manner: There is no certainty in will-transmission, especially in the will-transmission by language. Simply said, no language can guarantee the certainty of its own meaning.

Let us see how these ideas made their way into mathematics.

4.1. Being and Time Continuum

Postmodern thought is concerned with *discontinuity* and *difference*. Hence, it seems reasonable to start by considering what some mathematicians had to say about continuity (the continuum) and identity. It is always interesting to begin from the radical end of the spectrum, so first I consider some ideas of the twentieth-century Dutch mathematician L.E.J. Brouwer.

According to Brouwer, Kant confused the intuitive givenness of "the continuum" with the givenness of a particular geometry. He argued that with appropriate modifications, "the original interpretation of the continuum of Kant and Schopenhauer as pure a priori intuition can in essence be upheld."[9] But Brouwer's continuum has little to do with space as science represents it. The term has a special meaning for him, worthy of careful consideration.

For Brouwer, the choice of Euclidean geometry as the "geometry of the universe" is based on its practical value in our attempts to control the environment. I will return later to the idea that control is of some importance in Brouwer's philosophy, especially in his philosophy of language (sketchy as it was). For the time being, let us grant that we have no intuitive apprehension of the *geometry* of space, and that our theories of it are a matter of practical convenience impressed upon us by evolution, education, culture, desire for control, or something else.

The situation is different with the intuition of time. That we do have. Indeed, the intuition of time is for Brouwer the primordial intuition, the very basis of all conscious life. However, we have seriously damaged this primordial sense of time by treating it as we treat space, as if time were, so to speak, an additional dimension of that "practical" space that science peddles as the ultimate reality.

Brouwer concedes that our talk about "points in space" can be convenient and is acceptable as such. But it is also highly deceiving because it leads us to try to enumerate time in similar terms—as "points" and "moments," just as if we were observing physical objects—and thus to construct an inauthentic image of the *continuum* (of time). Our inner experience is fundamentally nonspatial and cannot be captured in scientific terms that we apply to space. This intuitive sense is a sense of the creative self unfolding freely in its private time.

This seems to be Bergson's idea, insofar as he applied some general romanticist notions to the particular case of time. Time as we conceptualize it, says Bergson, is measured time. Clocks, chronometers, digital watches, and other "objective" measuring instruments, split it up into atomic parts. Yet our sense of it is that of a continuous flux. That is why, in *Time and Free Will* (1898), Bergson introduced the difference between time and *duration*. The duration, says Bergson, "is something that exists *only* for us." It flows, it slides, it undulates, but it does not tick. It cannot be objectified or measured. It can be described, but no description can ever capture it fully. Lurking underneath this continuous flow is my real self, "the profound self" as Bergson says, a free human being that cannot be apprehended by reason, language, or measurement.[10] It cannot be seized, because the very imposition of form deforms it. As the Austrian writer Stephan Zweig wrote in his story "Buchmendel," it is "a sort of jellyfish glistening in the abysses of consciousness, slippery and unseizable."

This lack of conceptual framework to describe the private duration is indeed the source of freedom for Bergson. Our acts are free when they emanate from the entirety of our beings, when they are not stifled by the deterministic clockwork on which science is based. For Bergson, duration is a deeply personal experience that affirms individuality and ensures the possibility of freedom. To exist in this private "instant" is to be set apart. Our free acts in such instants possess a kind of indefinable resemblance to our real selves, a resemblance that, as Bergson wrote, "we sometimes see between a work and the artist."[11] Brouwer's intuitionism, in this respect, owes something to Bergson's intuitionism and thus to the entire romanticist tradition. However, Brouwer was considerably more radical than Bergson.

Let us examine how the construction of Brouwer's continuum proceeds. Brouwer describes the falling apart of a life-moment into a part that is passing away and a part that is becoming as "the fundamental phenomenon of the human mind" and in particular of mathematics. The recognition by the intellect of the possibility of indefinitely continuing a sequence of such "units" leads to the intuition of integers and fractions.

Here let me make a few remarks. First, note for future reference that Brouwer regards something that neither *is* nor *is-not*, but rather is a "falling apart," always between passing away and becoming, as the "primordial" feature of consciousness. Later, I will try to compare this idea to the French philosopher Jacques Derrida's "primordial and pure consciousness of *différance*," which also involves an ineffable falling apart. For now, keep in mind that this falling apart of a life-moment, this "bare two-oneness," is in particular the basic concept of all mathematics according to Brouwer.

Already here, although we are far still from the continuum, we can note a difference between Brouwer's construction and the "logocentric" one. Brouwer insists that numbers are the product of mental activities *of the individual*. This is clearly a point of disagreement. Most representatives of the "logocentric" tradition, from the German logician Gottlob Frege to Bertrand Russell and his co-author Alfred North Whitehead, regarded such "individualism" with a somewhat depreciative attitude. Russell objected to Kantian psychologism. Whitehead viewed his philosophy as the inverse of Kant's and held that "the subject emerges from the world." Against such views, Brouwer's position appears as an extremely radical individualism.

One or two more remarks, and then I will continue examining the continuum. For Brouwer, the inauthenticity of "logocentric" mathematics consists in its haughty disregard both for individual human beings and for what individuals can actually do, and in its replacement of the creative processes that go on in the human mind with abstract logical machinery that oversteps its legitimate application.

Unlike Kant, who wanted to construct a logic that does not "abstract from all contents of knowledge" but maintained that both kinds of logic are necessary even though they may end up in conflict, Brouwer dismisses abstract logic as a linguistic edifice that is to be classed with ethnography. His objection is especially directed at the treatment of infinity by simple extension of the methods that we apply to the finite, which thus leads us to paradoxes.

For this reason, Brouwer's views are sometimes related to the school of thought that admits only those mathematical objects that can be explicitly "constructed." That was part of Brouwer's agenda, but his ideas are somewhat different from those associated with "constructivist" schools of mathematics: Constructivists usually deal with quasi-mechanical processes of various kinds, while Brouwer rejected such notions altogether. He seems closer to what one might call romantic constructivism, which extends from Vico's "the truth is the made."

Let us return to Brouwer's notion of the continuum. To begin with, the continuum is not intuited as a collection of *points* that happen to be on anything like a straight line. It is a unified plurality that emerges from my realization that I could continue to insert numbers "between" those that I may already have constructed, to insert them not according to a determinate procedure or by giving them emptily conceived names, but rather by means of spontaneous, free, authentically individual choices. A way of thinking about the "points" of Brouwer's continuum would be to imagine them as open, indeterminate processes that actively involve the individual—creative processes that extend in the continuous privacy of inner time.

Such creative activities are for Brouwer "essentially languageless"—language is in some sense discrete, whereas the inner activity of the self is unfolding continuously—which would seem to make it impossible to try to give this flux a practical mathematical description.

It is a fairly standard objection to Bergson and Nietzsche that while they criticized the rigidity of words, they had to use language to describe what they preached could not be described. Mathematical language is supposed to be worse off in this respect. Even for Bergson, mathematics remains a construct of the analytic mind: It is the dead, immutable stuff of science. Brouwer takes exception to this and resolves the difficulty by introducing the concept of a choice-sequence (which he articulated around 1917).

A choice-sequence is a sequence of fractions. It is given by a finite initial segment together with a "rule" for continuing the sequence. The rule, however, is only partially determinate and involves making free choices. For instance, "choose a fraction that is greater than the preceding one in the sequence, but so that its square is less than 2." I can construct my sequence by choosing, at each step, the next element to be some (*any*) fraction that satisfies the condition above.

This is not a determinate process. The next fraction in my choice-sequence is any fraction that is greater than the preceding one, but such that its square is less than 2. The rule is deliberately "vague" so that at each step I can come up with as many fractions as I like that fit the rule. This indeterminacy leaves me the possibility of making free choices within the boundaries permissible by the rules. Each of these choices, even though it is guided by the rule, is still authentically mine.

It is important to observe that, unlike in the "classical" case where the construction from the preceding paragraph would lead to a limit-entity called "the square root of 2," Brouwer dispenses with the resulting limit-object and considers the construction *itself* as an "object." The square root of 2 has infinitely many decimals, and I am not in a position to know all of them. I can, of course, compute as many of its decimals as I like, but I cannot know the entire sequence. Hence, I cannot *know* this number as an object. For Brouwer, it is the *process* of construction that gets the status of an object, rather than the imagined "end product." However, the process could in general be different each time I perform the procedure, because it might involve spontaneous choices on my part.

Thus, this "object" will not be a garden-variety mathematical entity that I can fix once and for all. It is not immutable, because it depends on me. In a sense, the idea goes back to Kant. In the *Critique of Pure Reason*, Kant noted that "the number of parts [. . .] which a division may determine in a whole, will depend on how far we *care* to advance in the regress of the division."[12] Brouwer goes further and requires the process to involve making presently unknown free choices.

The continuum, then, is the unity of all such freely construed "objects": choice-sequences. It is a continuous, fluid paste from which points cannot be picked out with atomist accuracy. Its "atoms" cannot be apprehended because Brouwer's continuum is construed to flow with the flux of my own inner time, complete with its open future and the free choices that I cannot know now.

In a sense, Brouwer's continuum formalizes the idea of nonatomic time, expressed already by Schelling (and Bergson, later). Schelling's view of time as a "unity of division" was intended to account for the continuity of time:

> [Moments] can never be absolutely separated, so that never even for a moment could one be for itself. On the contrary, they are constantly posited as one *in* their separation, and it is precisely this that places them into the necessity of the process; for if they could be totally separated, then there would be no process.[13]

Summing up the effects of Brouwer's construction:

- For me, the "point" of the continuum is the active process of my consciousness taken together with the spontaneous choices I could make along the way.
- I can never permanently *fix* this "point," precisely because the construction of the sequence involves making free choices. The "point" of the continuum is not a standard mathematical object. It is not immutable. It is an *open* object, a construction with indeterminate future.
- The continuum cannot be split apart. I cannot pluck a single point out of it, a point I could call "now," because this point is an open object that depends on me. It does not wait for me to discover it, because I *create* it, freely, spontaneously, along with the plurality of all other "points."

Due to these and a few other things, intuitionist mathematicians must forgo the law of the excluded middle, which states that for every proposition P, either P or not-P is true. (This is the fundamental principle of the much-maligned "binary thinking" and "logic of identity," critiques of which are part of postmodern argumentation.)

But for two "points" of the continuum, X and Y, it cannot a priori be said that either "X is to the left of Y" or "X is not to the left of Y." X and Y are not atoms of space neatly arranged as infinitely small beads on an endless string. They are open objects; they live in time, just as you and I do. Attempts at a legitimate mathematical formulation of such a theory have been made, and have had other unusual consequences.

For example, there would be "points" that are neither inside nor outside a closed planar region, nor on the boundary of the region. Statements such that neither they nor their negations are true would become quite a common occurrence. To possess "truth" in intuitionist mathematics means to actually possess a mental construction called "proof." Since there are unresolved conjectures around, these conjectures are neither true nor false.

This is one of several important objections that Brouwer made. Employing the standard "binary logic," according to him, amounts to sneaking in the unwarranted assumption that every problem is solvable. Therefore: "The Principle of the Excluded Middle has only a scholastic and heuristic value, so that theorems that in their proof cannot avoid the use of this principle lack any mathematical content."[14]

That, among other reasons, might explain the lack of success of the intuitionist "revolution." Brouwer proposed a thorough revision of mathematical practices, and mathematicians tend to be a little touchy when faced with such serious revisionism. Many of the theorems people apply in engineering or business are not valid in intuitionist mathematics. Conversely, there are assertions that fail in classical mathematics but are provable in Brouwer's setup.

The specifics are not important here, but one can appreciate that this is a highly unorthodox species of mathematics. If we were to switch to it, we would have to reevaluate and try to reconstruct scores of things people had been taking for granted. On the other hand, this is part of what Brouwer's project was about: rethinking our approach to mathematics.

Let us now perform a thought-experiment to illustrate some curious nonmathematical effects of Brouwer's construction. First of all, let us observe that the self, for Brouwer, is not simply "in" time, but is fundamentally involved with it. Time is not some external dimension in which I live, have a lunch or two, and then die. Time, rather, is the very condition of all conscious life, the primordial intuition through which I conceive of everything. Being and time are somehow intertwined. As the Argentinian author Jorge Luis Borges wrote: "Time is a river that carries me along, but I am that river."

Imagine that I am a being thrown into the river of Brouwer's time continuum. I am not simply "in" it. It, too, is part of the fabric of my being, part of who I am. But who *am* I? *What* am I? It would seem that I cannot answer such a question because the answer—"I am . . ."—subtly involves me in singling out some object in the present, something that *is* now and will remain present long enough for me to think about it as a fixed object. I cannot do that in Brouwer's continuum. There is no "now" for me, only the "falling apart of a life-moment." There are no immutable objects that I can speak of, only the flow of consciousness of the "Creative Subject."

How, then, do I conceive of myself? It appears that I can do that only by casting a look forward, into the continuum of possibilities of what I *could* be in the future, into the cloud of my as of yet unknown free acts, the choice-sequences that *constitute* time for me. I am therefore always projecting, opening myself to the continuum of possibilities of what I could be.

So the result of my being thrown into Brouwer's time is this: I cannot know myself as I *am* because I conceive of the time continuum *itself* by "looking ahead." One way to describe this would be to say that I am not "present to myself." The ego is constantly pursuing the id—imagining and second-guessing its movements—but it never manages to apprehend it completely. (Where id *was*, says the famous Freudian tenet, ego *shall* be.)

Upon being thrown into Brouwer's continuum, I conceive of time (and myself) by looking ahead. This has something to do with Zeno's well-known paradox of Achilles and the tortoise, but perhaps we can compare the situation to something closer to our day. For example, the problem apparently found its way into Wittgenstein's *Philosophical Investigations*: "One is rushing ahead, and so cannot observe oneself rushing ahead."[15] By the time I describe something in language, it may already have become different. Hence I "compensate" for this tardiness by always anticipating how things could be. This perpetual looking ahead is reminiscent of something that Heidegger called "future-directedness" (*Vorlaufen*).

This could be interesting. Heidegger, on the one hand, is a philosopher who exerted a certain amount of influence on postmodern theory, while on the other hand his important work, *Being and Time* (1927), may have something to do with Brouwer's ideas about the continuum. Let us look at this briefly. (Heidegger expresses himself in a rather hermetic style, so I will try to formulate his point as painlessly as possible.)

Thrown into this strange world, the self is never fully "present" to itself, is never *in the present* for itself, but is a ceaseless opening toward the continuum of its existential possibilities. It is an open object, like other objects in the continuum. Thus, it is the structure of the continuum into which I am "thrown" that forces me always to be *concerned*, "future-directed," to *care* for my being. This seems rather close to Heidegger's definition of the subject (*Dasein*): the kind of being whose mode of being is to be concerned with its being.

It might also be interesting to note that Heidegger's critique of Bergson appears to be based on an elaboration of the situation encountered in our thought-experiment. Thinking the idea that the time is constituted by "looking ahead" through to its limit, it would seem that the past is also constituted by such a process. This leads to the apparently strange proposition that the past is constituted by a "looking ahead," so the past is "future-directed." Perhaps this could be understood in the following way. One thinks of the past not only as it *was*—in a crudely empirical "core dump" of the contents of our memories, a procedure that Heidegger attributes to Bergson—but also as it *could* have been. Thus, the continuity of our "lost time" is constituted by intertwining empirical recollections with interpretive actions (which are bounded but not entirely determined by what one can literally recollect).

Let us pursue this "Heidegger connection" a little further. Brouwer's critique of classical mathematics, which treats infinity as if it were a finite collection of objects always at our disposal, could in Heidegger's terms be translated as the critique of all thought that does not admit its own finitude (its "being-for-death") and proceeds in terms of some inauthentic, anonymous "they"—the "they" who, unlike the finite and mortal I, are supposed to be able to stretch themselves into the infinite future, a time line in the standard abstract sense. "They" are the people who think they can *know* the square root of 2 because they can describe it in mathematical language and approximate it with any desired precision.

Furthermore, like Brouwer, who spoke of "mathematical attention" as a fundamental human activity that slips away from the full grasp of logic, Heidegger speaks of "the mathematical" in what seem to be similar terms. To begin with, he seems to have distinguished between "the mathematical" and "mathematical formalization." In the essay "Modern science, metaphysics, and mathematics," Heidegger wrote: "The mathematical is that evident aspect of things within which we are always already moving [. . .]. Therefore, the mathematical is the fundamental presupposition of the knowledge of things."[16]

This "deep structure" provides, as it were, a context where meaning arises. It always mediates in our understanding of the world. Nonetheless, this context, "the mathematical," in a sense eludes complete formalization: "The phenomenal context of these 'Relations' and 'Relata,'" he says in *Being and Time*, "is such that they resist mathematical formalization."[17]

Both Brouwer and Heidegger have some strong words for science and technology—Heidegger is known for having nightmares about technology—and for abstract logic. "By continually appealing to the logical," wrote Heidegger, "one conjures up the illusion that he is entering straightforwardly into thinking when in fact he has disavowed it."[18] Below I discuss how well this fits in with some of Brouwer's views.

But such similarities go only so far. There are crucial differences, too, and I point them out to avoid getting carried away and stretching the above analogies into the realm of the utterly inauthentic.

For example, Brouwer means to liberate us from the belief that language exercises a magic power over the individual, while Heidegger, at least on *some* readings, ends up claiming that it is language, ultimately, that speaks *itself* through the individual. As opposed to Brouwer's extreme assertions about mathematics as an "essentially languageless activity," Heidegger pointed out both the necessity of linguistic formalization and the impossibility of ultimate formalistic reductions of "the mathematical." Whereas Brouwer insisted that truth is a process of creation, a sovereign construct of the mysterious "Creative Subject," Heidegger maintained that it is a process of revelation, disclosure, or "unconcealment."

Extrapolating from Heidegger's ideas about truth in the work of art, one may say that truth in the work of mathematics is disclosed through mathematical activity, becomes a familiar, reliable part of our world, and is assimilated into our "intuitive" understanding. (So intuition loses its primacy, but it still plays role in the "vicious circle" through which the intuitive and the formal-linguistic aspects influence one another.)

Here is another point of divergence. Brouwer believed that the continuum is a construct of consciousness, while Heidegger seems to have regarded it as an unconscious "place" where consciousness emerges, a "place," furthermore, that has been interpreted in different ways throughout history. Brouwer's self-centered idealism is thus checked by Heidegger's assertions to the effect that the subject is only the guardian of truth, its "shepherd" but not its "master."[19]

These are crucial disagreements. We should definitely keep them in mind, because it seems that the differences between intuitionist mathematics and *some* aspects of postmodern theory are analogous to these differences between Brouwer and Heidegger. Nevertheless, we should not overlook the empirical fact that a mathematical pattern on the same pair of socks can look deceivingly different when the socks are turned inside out. Since I will eventually compare intuitionist argumentative methods to those of postmodern theory, and since Heidegger is such an important precursor of that particular enterprise, let me also point out the possibility of more concrete historical connections.

It is not completely inconceivable that this theory of the time continuum, which was causing something of a "scandal" in the decade that led up to *Being and Time*, would have been an issue about which Heidegger cared.[20] Heidegger was, after all, a student of mathematics as well as of philosophy. For instance, one of his earliest publications, "New Investigations in Logic" (1912), assesses the work of Frege, Russell, and Whitehead. (The first installment of Russell and Whitehead's impossibly formalistic masterpiece *Principia Mathematica* appeared in 1910.) Finally, Heidegger was deeply influenced by Edmund Husserl, himself a mathematically trained philosopher, whose connections with intuitionism we have yet to explore.

I will pick up this line of thought later. For now, let me simply note the possibility of a connection—even if only a limited one—between a mathematician and such a pillar of postmodern theory as is Heidegger.

4.2. Language and Will

Let us look into what Brouwer has to say about knowledge and language. Mathematical knowledge, for Brouwer, is always an active determining, the work of the Creative Subject that unfolds "in its deepest home," far from all speaking and reasoning. Mathematics is a process of creation. It is "knowledge" in roughly the same sense that knowledge of the self is "knowledge" for some of the romanticists. The self is felt, experienced, lived. It is never given with logical certainty because, unlike logical certainties, it is not meant to be around forever. It is (among other things) a living and a mortal body. Self-knowledge has no *logical* basis, and cannot have such a basis, other than in the sense of a prelinguistic familiarity.

Mathematics is a way of realizing the creative potential of the self, which comes to discern itself through doing mathematics, just like the romanticist selves realize themselves through art. Indeed, Brouwer picks up the familiar theme of the "liberation of the mind" (a favorite phrase of his early treatise *Life, Art, and Mysticism* [1905]). He equates beauty, truth, and goodness, and regards all acts of "cunning" or "calculation," especially as exemplified in science and "social acting," to be unbeautiful, which is to say that they are in some sense morally impugnable.

It is difficult to place Brouwer within a single philosophical tradition. Yet he seems to show some affinity with romanticism in a broad sense. Brouwer, for instance, professes (in *Life, Art, and Mysticism*) that when all the pathetic philosophical attempts to fill the gaps in our understanding have finally failed, the wiser among us might hold on to the notion of the ego that comprehends but is itself beyond comprehension. However, he then proceeds to describe that approach as the ultimate philosophical "plug," which could be construed as a jab at romantic idealism.

But such a subtle reprimand—calling something a philosophical "plug"— sounds more like a term of endearment when compared to Brouwer's otherwise extremely bombastic style. Consider what he thinks of those who might choose a different philosophical plug. These unfortunate people, scientists especially, who look for some final and ultimate certainty "keep on and on until they go mad; they grow bald, short-sighted and fat, their stomachs stop working properly, and moaning with asthma and gastric trouble, they fancy that in this way the equilibrium is within reach and almost reached."[21]

Science represents, as it generally does for the less restrained romanticists, an ossification, "death." For Brouwer, science is an "infatuation of desire restricted to the human mind." It represents a symbolic Fall of Man caused by the intellect, "that gift of the Devil." Logic and science are "to be classed with ethnography." Logic is particularly not the basis of mathematics. If anything, logic and science are a degenerate form of mathematics. They are mere records of some mathematical activity that is essentially beyond language. There is nothing wrong, says Brouwer, with studying logic and science, as long as we know that we are thus studying how people organize their thoughts. (That is why logic is a kind of "ethnography" for him.)

In this sense, and only in this sense, science and logic are acceptable. But interpreting mathematics as being *based* on logic is "like considering the human body to be an application of the science of anatomy."[22]

This is an important issue for Brouwer. Mathematics is an act of will, of creation, while language is at best a flawed vehicle for the transmission of that will. "Linguistic edifices, sequences of sentences that follow one another according to the laws of logic," says Brouwer in *On the Foundations of Mathematics* (1907), "have nothing to do with mathematics, which is outside of this edifice."[23] Elsewhere he added, quite controversially:

> Intuitionist mathematics should be completely separated from mathematical language and hence from the language of theoretical logic, recognizing that intuitionist mathematics is an essentially languageless activity of the mind having its origin in the perception of a move of time.[24]

This sounds unnecessarily radical. But we have to understand that mathematics is for Brouwer what art was for every romanticist: a continuous creative flow, "free will," some inner activity of the mind that cannot be reduced to, or deduced from, language. So it would be better to say that it *eludes* language than that it is "essentially languageless." This is consistent with romanticism in general, and with other Brouwer's claims in particular. In a lecture delivered in Vienna in 1928, Brouwer said: "There is neither exactness nor certainty in will transmission, especially not in will transmission by language [. . .]. There is, therefore, also for pure mathematics *no certain language* [. . .]."[25]

One might be tempted to hear an echo of Nietzsche in all these incantations of the will. Everything is a creation for Nietzsche, a will-to-overpower disguised as a will for truth, and there is no way to categorize such "processes ad infinitum" by means of words. There is no ultimate reality for Nietzsche. Language cannot hold on to anything permanently without an act of will. "We have arranged for ourselves a world in which we can live—with the acceptance of bodies, lines, surfaces," says Nietzsche. "Without these articles of faith no one now would be able to live. But this by no means constitutes a proof. Life is no argument. Among the conditions of life, error might be one."[26]

We see things in a certain manner, and *then* perform logical operations on them. Logic, on its own, does not offer any image of the world. The world that logic and science display is not the objective world. It rests on a previous interpretation. Therefore, according to Nietzsche, the apparent world is the only one; the "real world" is merely a lie.

For Brouwer, too, the "real world" is an apparition. He even goes as far as to say that my "exterior world" consists, for me, of "things" that are essentially only sequences of my own thoughts, such that even the existence of other minds is a "mere hypothesis" (a significant departure from the early romantics). The world's perceived form is essentially the act of the will, in some rather general sense of that word. This act of "mathematical attention" precedes all logic:

> A special case [. . .] is the construction of *objects* in thought, that is, of persistent, permanent things (simple or compound) of the perceptual world, so that at the same time the perceptual world becomes stabilized. [These] phases of mathematical attention are in no way merely passive attitudes; on the contrary, they are the acts of the will.[27]

But unlike Nietzsche, who is sometimes taken as arguing that culture and language ("grammar") are imprinted into the self and even into the body by a process of "mnemotechnical" inscription—a branding so inescapable that it takes a Superhuman to break out of the "herd"—Brouwer maintains that it does not take a superhuman effort at all. There is no need to call for a Dionysian frenzy of power and desire *à la* Nietzsche. Mathematics, arguably the last place where one would look for a "subversion," transgresses the straightjacket of language.

The creative-interpretive acts of the individual mathematician who lives in his linguistically ineffable continuum—let me use the term that Schleiermacher used in a similar context—cannot be "mechanized." They can be linguistically motivated, guided by a rule, or rationalized a posteriori, but they remain free.

In this, at least, Brouwer seems closer to early forms of romanticism than to the poetic flights of Nietzsche, with whom he nevertheless shares a penchant for being shocking. Thus, when Brouwer claims that no one had ever been soul-to-soul with anyone else, meaning that no one had ever unambiguously known what someone else means, it appears that one of the possible contexts of such a viewpoint may be found in, say, Humboldt's view that no one means the same thing as someone else.

Let us look at Brouwer's ideas on language in more detail; they will become relevant later, in the discussion of Wittgenstein.

The acts of individual will—mathematics included—first "served the individual man." But they can be made to serve others "in the form of *labor*." This can be done directly, either by "suggestion," which for Brouwer means "striking terror," or by "seduction, by stirring the imagination," or it can be done indirectly, by "training the mind, that is, by influencing the experience of the person to be enslaved in such a way that he adopts a [...] view in which the prospect of pleasure or the fear of pain produces a readiness to work."[28]

But the method of training the mind, says Brouwer, is wholly inadequate to ensure loyalty of the individual. Societies thus resort to the "propaganda of moral theories" and advertize values that transcend the egotistic attitudes. This cannot work either, because the egotistic individual can use moral values as a means of restraining the ambitions of others.

What remains, in terms of seductive devices, is language itself. In the primitive societies, simple gestures or a single human cry may have sufficed for the transmission of will. But as social organization grows more complex, tasks become too diverse to be induced by a single human cry. Due to this growing complexity, language evolves as a subtle method of imposing orders on the entire society. Unlike moral "propaganda," language is necessary for communication and can easily be advertized as being morally neutral.

Therefore, language itself turns into a subtle form of ideology. It is given the status of the objective carrier of meaning—much like science grants geometric space the status of objective world, from Galileo onward—by subjecting it to mathematics-like rules that are themselves supposed to be unquestionable. Then orders can be issued to armies at once, provided that all individuals are trained to believe that language *itself* has some ultimate and unambiguous meaning that transcends all individuality. In other words, to prevent the egotistic individuals from perverting the meaning of language and turning it to their own advantage, a belief must be instilled in them that language has some unambiguous meaning all of its own. This is what Brouwer calls the false belief in the magical character of language.

Brouwer is out to fight this false belief. Mathematics, being the creative act of individual will, cannot be completely reduced to language—this is the central point that Brouwer makes—and it follows that language is only an aid, something that makes social organization *possible*: "In this way most of the will transmissions required by civilized society are made possible. Language, therefore, is altogether a function of the activity of social man."[29]

Meaning does not magically spring out from language itself. The meaning of language always involves a consensus, achieved by training and "seduction" and enforced by other methods available to society. However, nothing can guarantee the universal loyalty of the individual to any such consensus. You can make me act according to some rule, train me, impose it on me, and I may follow it if I am sufficiently frightened or otherwise "seduced." But you cannot take away the spontaneity of my interpretations of the rule. I *myself* cannot wipe out my ability to reinterpret it, although I might be forced to suspend it for pragmatic reasons: I need to function as a social being.

Thus, the processes of socialization and culturalization may well put in question my absolute autonomy as an individual. This is not inconsistent with the romantic "philosophy of finitude," which already saw the self as dependent on something that it cannot fully comprehend. For romanticism, the individual cannot be entirely subjected to communal linguistic practices precisely *because* the mysterious creative force within her is beyond language. Brouwer is close to this viewpoint: "[A]ll speaking and reasoning is an attention at a great distance from the Self; we cannot even get near it by reasoning and words."[30] Let me sum it up in a less radical manner: The social-cultural context is *formative*, but it is no way *determining*.

It could have been this stubborn privacy of Brouwer's "free will" that intrigued Wittgenstein—on hearing Brouwer's lectures in Vienna in 1928—and led him to devise one of his celebrated arguments on the indeterminacy of rules. For instance, when Wittgenstein wrote that "no course of action could be determined by a rule" and gave a *mathematical* rule as an example (in addition to many other examples, of course), it may be said that he is partly transmitting Brouwer's will. "There is [. . .] for pure mathematics *no certain language*," wrote Brouwer, because no rule, linguistic or logical, can transmit the will with absolute certainty.[31] Language is not magical.

Wittgenstein's famous argument against the possibility of a "private language"—which I discuss in detail later—also seems related to Brouwer's (and Nietzsche's) notion that language is "altogether a function of social man." Finally, when Wittgenstein discusses exclamations such as "Sit!" it seems to be an echo of Brouwer's "single human cry."

As opposed to some of his Viennese and Cambridge friends, Wittgenstein was willing to reexamine his early "logically positive" viewpoint. In confronting Brouwer's outrageously eccentric pronouncements about mathematics as an "essentially languageless activity," he constructed an important (though controversial and only slightly less eccentric) philosophical argument. And he based part of it on a surprisingly simple mathematical trick.

This is not, of course, to say that he simply *agreed* with Brouwer. Yet Brouwer seems to have been a formative influence on Wittgenstein, which suffices for my purposes of establishing the possibility of a "postmodern connection."

But let me postpone stepping into the minefield of interpreting Wittgenstein's utterances until later. For the time being, let me sum up the possibilities I have indicated in this chapter: Brouwer's ideas about language (mathematical language in particular) and the time continuum could be seen as connected to—but not identical with—certain views of each of the Three Wise Men of the postmodern vulgate: Heidegger, Nietzsche, Wittgenstein.

5
BEYOND THE CODE

Brouwer was not the only mathematician who objected to the "atomistic" view of space and time, and to the limitations of a linguistically obsessed view of mathematics. German mathematician Hermann Weyl made similar objections, as well as Henri Poincaré and several other French mathematicians to whom Brouwer referred as "pre-intuitionists."

It is important to note that none of them denied the *practical* value of the atomist view, including the idea of time as a "line" filled with points-moments that are in essence regarded as linguistically accessible entities. This mathematical idealization proved itself an invaluable methodological tool for modeling physical phenomena, motion in particular. That was not under dispute at all. What came under fire was the deeper issue of whether this atomic theory is satisfactory when it presumes to offer objective knowledge of the continuum.

The general strategy of these critiques seems to have been to outline some "paranormal" consequences of the assumption that linguistic devices can capture the sense of continuity, with the intention of demonstrating that denying some form of intuition is not a plausible approach. In this sense, the continuum consistently appeared as something other-than-language, something that can be given only by an extralinguistic capacity of the self. Here are a few examples.

This is an interesting excerpt from Weyl's 1925 article "The Current Epistemological Situation in Mathematics":

> And if it were really possible, in accordance to Zeno's paradox, to put together the line segment of length 1 out of the infinitely many subsegments of length ½, ¼, ⅛, . . . , qua "hacked off" wholes, then it would not be understandable why a machine, if it manages to run through all these infi-

nitely many segments in a finite time, could not also carry out in a finite time an infinite sequence of distinct acts of decision, say, by producing the first result after ½ minute, the second ¼ minute later, the third ⅛ minute after the second, etc.[1]

There might be practical difficulties with this idea, but some people (the original "hackers") have been intrigued by the possibility. However, under the natural assumption that there are some physical bounds to computational speeds, Weyl's remark reveals a relationship between the continuum and the limits of computability. Indeed, British mathematician Alan Turing proved his celebrated undecidability theorem in the 1930s precisely by considering uncomputable "points" of the continuum. We should take a quick glance at this important result.

Turing's argument proceeds roughly as follows. Suppose we have a "generic computer" at our disposal. We are looking at the possibility of generating the atoms of the continuum by writing various programs that produce, say, numbers between 0 and 1 in their decimal expansion. Let us be generous and list *all* syntactically valid programs. Denote them $P(1)$, $P(2)$, and so on. The first program $P(1)$ outputs the first number, that is, the sequence of digits that constitute its decimal expansion. Denote these digits by $P(1,1)$, $P(1,2)$, and so on. Since the program $P(1)$ is "any" syntactically valid program, and since some valid programs can end up running in infinite loops, it is possible that there is no output $P(1,n)$ for some n. We will ignore that for now.

The second program $P(2)$ produces the second atom of the continuum, with the decimals $P(2,1)$, $P(2,2)$, and so forth. As before, there is a chance that some of these computations will "draw a blank," a possibility that we will ignore for the time being. Continuing in this way, we will have listed all the programs that compute anything. So all the computable points of the continuum must appear on the list of outputs of these programs.

Now define a point of the continuum as follows. Its first decimal is any digit other than $P(1,1)$, its second digit is any digit other than $P(2,2)$, and, in general, its nth decimal is anything but $P(n,n)$. It seems that this number is computable. I could, for instance, write a program Q that for each n generates the program $P(n)$, computes $P(n,n)$, and changes its output to anything else. The number that Q would output would be different from all the numbers that the programs $P(n)$ compute, and hence it could not be on the list of computable numbers.

This is absurd: I just wrote a program that computes it. I must have made a mistake. The mistake, as usual, is hidden in the parts that I said we would ignore.

We have to deal with the possibility that some of the computations $P(n,m)$ might "draw a blank." Some of them might run in a loop forever, without stopping and outputting a result. Hence, my program Q would not "know" how to change the digit I want it to change—there may be *no* digit on the way, and it would wait in vain. The problem would be "resolved" if Q could call up a subroutine that decides whether any computation in any given syntactically valid program halts or not. Then I would actually be able to write Q, and I would truly be in trouble. I would have a program that computes a number that by its very definition cannot be computed by any program. The only way out is to concede that there is *no* program that is able to decide which programs stop and which do not. The problem is "undecidable."

This result of Turing's is known as "the undecidability of the Halting Problem." It raises interesting questions regarding the limits of formal reasoning, but for now I would simply like to note that the problem of the continuum leads, fairly directly, to an important theorem in the theory of computability. The continuum, so to speak, is "the other of any programming language."

Let us look at another example, one that invokes the French language. Emile Borel, a well-known French mathematician, joined the discussion of the continuum in the 1920s, on the side of Brouwer and Weyl. Consider Borel's 1927 note to *Revue de Métaphysique et de Morale*. Borel observes that *knowledge* of every point of the continuum—the "knowledge" that the atomist takes for granted simply because he or she can define that point, linguistically hack it off by some "singular description"—would have some strange consequences. Here is what he says (I edited the text slightly):

> One could define [a] number by saying that each of [the] successive digits [of its decimal expansion] is equal to 0 or 1 according to whether the answer to some question or other is affirmative or negative. Moreover, it would be possible to order all the questions that can be asked in the French language by sorting them [. . .] as is done in dictionaries. Only those questions for which the answer is *yes* or *no* would be retained. The mere knowledge of the number thus defined would give answers to all past, present, and future enigmas of science, history, and curiosity.[2]

From this point of view, claiming anything like the actual knowledge of such a logically definable "atom" of the continuum—knowledge of this point as a separate entity—seems a bit optimistic. Yet it is a consequence of the "logocentric" view of mathematics that we should have such knowledge. We have the knowledge of logic. Classical mathematics, and in particular the theory of the continuum as a collection of nameable "atoms," is supposed to be reducible to logic. Hence, we should know Borel's number, which is of course somewhat unlikely.

I mention this idea for two reasons: first, because it can be made into a result about "randomness in mathematics," and second, because I believe it illustrates the general strategy of exhibiting the continuum as "the other of language"—a kind of flux, an incessant flow that cannot be immutably categorized by words or permanently "hacked off" into atomic bits. This "other" will become important later, when we consider Derrida's notion of *différance*, which also appears to be "the other of language"—I have heard people speak of the "unspeakable *différance*."

5.1. Medium of Free Becoming

The above digression aside, I would like to concentrate on the role of Weyl, an important figure in the scientific circles in the first half of the twentieth century. In addition to his contributions to mathematics and mathematical physics—which are not small by any standard, though Weyl does not appear in television commercials nearly as often as Einstein—he maintained a keen interest in philosophy. In one way or another, Weyl, too, entertained some "romanticist" ideas. For example, in his articles from the 1920s, he refers to Fichte and quotes from Nietzsche.

What is interesting for us here is a possible exchange of ideas between Weyl and Husserl. Before turning to philosophy, Husserl studied mathematics under some of the most prominent mathematicians of the time, and he always maintained an interest in it. His philosophical outlook, phenomenology, was in a sense an attempt to construct an exact scientific study of that which is given to consciousness, a kind of "mathematics of phenomena." He and Weyl knew each other from when they were both in Göttingen, and they remained in contact after Husserl moved to Freiburg and Weyl to Zürich.

Weyl outlined a critique of the standard theory of the continuum already in his *Das Kontinuum* (1918), which Husserl read and apparently admired. "Finally a mathematician," he wrote in a letter to Weyl, commenting on *Das Kontinuum*, "who understands the necessity of phenomenological thinking [and of] finding a way toward the primal ground of logical-mathematical intuition [. . .]."[3]

In some of his work from the 1920s, Weyl clarifies and further develops Brouwer's ideas in his own way. Instead of Brouwer's choice-sequences, Weyl looks at double-sequences of fractions. For example, I can imagine a sequence of nested intervals like (1,2), (1.1,1.9), (1.11,1.89), and so on. The idea is that this sequence of nested intervals tells us "where" the "point" of the continuum that it represents should be. Each interval is contained in the preceding one, so this process of "subdivision" becomes more precise as I go along.

But it is not the final result that we are after. As with Brouwer's choice-sequences, it is the process that matters. In such a process, I am allowed to come up with the sequences that demarcate the boundaries of the intervals according to a rule, but in addition to that, I am also allowed to choose the endpoints *freely* (as long as I do it in such a way that the resulting intervals remain nested).

I think we can safely suppress other technical intricacies and retain this much: Weyl models the continuum in terms of double-sequences. This "doubling up" will play an interesting role later on. Also note that Weyl, at least for a while, supported Brouwer's idea of the intuitive givenness of the continuum as a construction involving free individual choices. For example, here is an excerpt from "On the new foundational crisis of mathematics," an article he published in 1921:

> Brouwer's remark is simple but deep: we have here the creation of the "continuum," which, although containing individual real numbers, does not dissolve into a set of real numbers as finished beings; we rather have a *medium of free Becoming*. We found ourselves in the domain of an age-old problem of thought, the problem of continuity, of change, and of Becoming.[4]

The new theory of the continuum, Weyl comments, is an undertaking intended "to do justice to *Becoming* in a valid and tenable manner." (Weyl later retracted some of his more enthusiastic exclamations, such as "Brouwer—that is the revolution!" The assertion had prompted a certain Cambridge fellow to call Brouwer and Weyl "Bolsheviks." Wittgenstein came to their defense, referring to the fellow as a "bourgeois philosopher.")

Let me try to point out why all this may be relevant to Husserl's phenomenology. At first glance, the possibility of a connection suggests itself in Husserl's earliest works. For example, in *On the Concept of Number* (1887) and *Philosophy of Arithmetic* (1891)—written while Brouwer was a very young lad—Husserl spoke of "acts of consciousness" and "mental constructions." In fact, one of Husserl's mathematics professors, the "constructivist" Leopold Kronecker, was in a sense a precursor of intuitionism. Yet unlike Brouwer, who was interested in acts of construction *themselves*, Husserl wanted to provide an objective *description* of what is given to consciousness through such acts. So there are important differences between intuitionism and phenomenology. But let us keep looking for what could be construed as points of convergence.

Husserl's method consists in a systematic bracketing off ("reduction") of all experiences of the external world, in order to clear the room for the appearance of that which is given in pure intuition. By the end of this process, Husserl hopes to arrive at the evidence of the presence of transcendental ego, a kind of "generic" self that appears to itself in purely intuitive light. It is the "principle of principles," the Archimedean point from which all knowledge derives its security. The study of this self-presence would then form a basis for a "mathematics of phenomena," an exact science of how things are construed by consciousness.

Such a reduction could potentially be problematic when it comes to language. It is not immediately clear how the evidence of anything could be collected in a cognizant manner if language were finally bracketed off. If I *know* something, then I should be able to justify it, and that involves language. Yet language is a social phenomenon, and for that reason it might have to be bracketed away (unless there is something like a "private language," which is a problem in itself).

Husserl attempted to resolve this communication problem by introducing (in his *later* work) the concept of a communal "life-world," the world of prescientific everyday experiences and mental activities. Practical individual experiences are of course different and diverse. Husserl nevertheless maintained that there is a fundamental "core" of our life-world that is equally accessible to everyone.

This is not the idealized world of immutable objects with which classical mathematics deals, nor is it the "Kantian" world of universally formed *sense*-experiences. It is a "medium" where objects are constructed in consciousness, in their continually changing aspects. I am now tempted to say that the medium of which Husserl wrote is none other than Weyl's "medium of free becoming." In fact, in his *Cartesian Meditations* (1929), Husserl quite explicitly takes up the idea that objects are always construed not as they *are*, but as they *could* be: Our experiences of physical objects are "an open, infinite, indeterminately general horizon, comprising what is itself not strictly perceived—a horizon [. . .] that can be opened up by possible experiences."[5] However, Husserl seems to have come upon his version of the "medium of free becoming" independently from Brouwer and Weyl.

Although the concept of the life-world is officially introduced only later, already in the 1911 lecture "Philosophy as a rigorous science," Husserl spoke of the "flow of phenomena" that are not divisible into components and are not "analyzable in the proper sense." Whatever this medium is, it is, according to Husserl, the "original ground" without which no knowledge is possible, the crucial element suppressed in scientific formalizations ever since Galileo turned the universe into a geometric idealization. So this life-world shares some properties of that which Heidegger called "the mathematical." Its core rests in the constitution of time-consciousness.

In "Philosophy as a rigorous science" Husserl clearly indicates just what *kind* of time he has in mind: "It is [. . .] a time that no chronometer can measure."[6] In a sense this core is something *like* Brouwer's time. Husserl, so to speak, turned the private continuum into public domain.

I think we can now take a safe guess where the problems begin. Husserl has to justify that this core of the life-world is always "present" and remains unchanged throughout history. It is not enough, for his purposes, that *he* can construe the continuum in his consciousness. Husserl must demonstrate that everyone—from Adam to Captain Picard and beyond—*necessarily* construes the continuum in the same way, that the constitution of our sense of time proceeds according to a timelessly valid form.

Otherwise, mathematics would be a historically conditioned kind of knowledge. The primordial intuition would remain beyond the reach of knowledge, and its timeless presence could not be guaranteed. Mathematics would then not be the rationally grounded, historically unified, universal construct of sovereign thinking-subjects. It would be grounded in a series of historically contingent intuitive images "tendered" to us by the grace of I-know-not-what—just as it is for Heidegger, who christened this mysterious process by introducing the even more mystifying coinage "the clearing of Being." This issue, in a few steps that I outline below, brings Husserl right back to the problem of logical justification of continuity.

Husserl began to struggle with the notorious "other of language" already by 1905, as he was completing *Lectures on the Phenomenology of Internal Time-Consciousness*. (The text was edited and prepared for publication by Heidegger, in 1927.) There, he was led to conceive of time-consciousness as a *duality* of moments precisely in order to account for the continuity of our sense of time. Time, somehow, is "two dimensional." One naturally wonders what that could possibly mean. The notion seems not a little strange.

The idea at first seems related to Weyl's modeling of the continuum by means of double-sequences. But Weyl's construction came after Husserl's, so let us try to imagine in less technical terms what Husserl might have meant by "two-dimensionality" of time-consciousness. On the level of common sense and rough analogy, it is perhaps not so difficult to illustrate it.

Suppose, for example, that this section of my book were to end abruptly right here, say, by quoting an entire short story by the Guatemalan writer Augusto Monterroso: "And when I woke up, the dinosaur was still there." You would presumably be slightly surprised, for at least *two* reasons: first, because you would have retained enough of the preceding text to register the one-sentence story as a "discontinuity," or at least as a more violent digression than you had seen so far; second, because while reading the preceding text you were looking ahead—"protending," says Husserl—and so "predicting" the future course of the text. You performed some imaginative variations, however hazy and indeterminate, about what *could* come along.

Loosely speaking, there seem to be two "degrees of freedom" in the construction of such a sense of continuity. You are *a priori* (and for no good reason) presupposing some "continuity" of the text. But since you do not know what will come next, "continuity" can only be observed *a posteriori* (based on a retroactive comparison with what came before).

Husserl was more ambitious and scientifically minded than were Brouwer and Heidegger, so to ensure the universal and timeless validity of his "mathematics of phenomena," he had to explain how he would *know* that the intuition of time continuum remains invariant throughout history. Hence, Husserl is forced to treat the intuitively constructed continuum as an immutable object of formal-logical knowledge, which he can somehow observe as from the "outside." Phenomenology doesn't seem likely to provide that much.

Let me sum up what seems to be the difficulty. Husserl can justify the claim that he has actual knowledge of the continuum—the knowledge that time-consciousness is constituted in a historically invariant manner—only by ascending to some point *beyond* time, beyond history. He is aware of this and eventually invokes the total flux of lived experience, of which one can at best say that it is an idea "in the Kantian sense." However, like other Kantian ideas about totalities, this idea is not knowledge.

Derrida chooses precisely this point to carry out part of his critique of Husserl in the "introduction" to *Origins of Geometry*. (The introduction is considerably longer than Husserl's original text.) I cannot hop outside of time at my personal convenience in order to verify that my sense of time is invariant in time. My observations about how time-consciousness is constituted may be historically contingent. Hence, I cannot fully grasp the "origin of geometry" and thus finally bring it whole into the luminous circle of my insight. I am a shepherd of truth, not its master.

In the language of postmodern theory—which is fond of such metaphors—one might say that the *duality* ("difference") involved in the constitution of the continuum cannot be brought into a knowable *unity* ("identity").

Here is what Derrida actually says in his introduction: "This thought unity, which makes the phenomenalization of time possible, is therefore always an Idea in the Kantian sense which never phenomenolizes itself." And we are aware of it too: "[T]his impotence and this impossibility are given in a primordial and pure consciousness of Difference."[7]

If we suppress metaphors with which Derrida's arguments are teeming, it may begin to seem that his objection to Husserl basically goes back to Heraclitus's classical riddle (recall from chapter 2): What persists through all changes? The answer: change itself. In a similar vein, Husserl, who actually uses the expression "Heraclitean flux," wants to say that it is this flux itself that persists through all changes. Derrida simply points out that Husserl cannot *know* that, unless he steps out of the whole fluctuation business, that is, history. Hence, the life-world, for all we know, is a historical thing through and through: It cannot be guaranteed to have an immutable core.

Very well, but how is this different from Heidegger? Did he not already historicize that in which "we always already move," the world known to us through "the mathematical"? There are certain similarities between Derrida and Heidegger. But there are also crucial differences, not least in their methods. For one, Derrida has the entire quasi-mathematical arsenal of continental structuralism at his disposal as his immediate historical precursor and the sublime object of his critique.

I find this very exciting, because it makes Derrida's work even more accessible to a critical reading from a mathematical standpoint. So far we have merely glanced, almost incidentally, at one of his less popular publications. Later, when comparing Derrida's notion of *différance* to the notion of the continuum, we will find that he uses an expression similar to Weyl's "medium of free becoming" to describe this pure difference that causes Husserl's principle of principles to slip away (or, as Derrida says, to be "deferred-delayed without respite"). I will try to show that this intuitionist connection offers an interesting context for postmodern theory's attacks on "binary logic" and "logic of identity."

Once we have looked at the language-theoretic background of continental philosophy, we will be in a position to determine in what sense it could be said that Derrida relies on, reinvents, or transforms parts of the intuitionist critique of "logocentric" mathematics (even though his philosophical outlook is very far from Brouwer's idealism). We have a lot of ground to cover before then. What I am trying to indicate at this stage is that mathematics may have had its hand in the arguments of Derrida's dialogical partners. Husserl and Heidegger are certainly among them. Poincaré does not typically appear on that list—but he could, and perhaps should.

5.2. Nonpresence of Identity

Poincaré, like Brouwer and Weyl, defended the view that mathematical inference is different from formal-logical inference. Most interesting for the purposes of unearthing some "postmodern" connections is Poincaré's concern with how the identity of objects is given and with how the notion of continuity comes about. But let us first see how we could situate Poincaré in an attitudinal vicinity of romanticist critiques of science.

Poincaré's campaign against Russell's and Frege's logicization of mathematics may be a good place to start. It is not essential to go into the details of their exchange, although some of it is rather entertaining. For instance, when Russell discovered a certain paradox in Frege's system in 1901, Poincaré could not quite contain his pleasure: "*Logistique* has finally proved that it is not completely sterile. At last it has given birth—to a contradiction."[8] Here we can see part of Poincaré's objection: Mathematics cannot be equated with logic; unlike logic, it is not "sterile." Mathematical understanding is in a sense creative.

Russell and Frege maintained that mathematics is reducible to logical inference, that it is detached from its human practitioner insofar as one only needs logically to analyze mathematical text—"scan" two propositions in order to show that one implies the other—to infer its meaning. On this view, even the most mundane logical inferences are to be considered productive (i.e., leading to new knowledge, "creative") as long as they allow us to produce something *formally* different from the premises.

Poincaré reacted sharply to such a view. The logical reasoner's grasp of mathematics, says Poincaré in one of the many comparisons of this kind, is like the understanding of elephanthood that someone gains upon microscopic examination of elephant tissue; the difference between mathematical and logical inference is like the difference between a writer who knows the rules of grammar and one who has a story to tell; reducing mathematics to logic is like reducing chess to the rules for moving pieces on the board.[9] Mathematical understanding is not something that can be completely captured by the ("grammatical") rules; there is always an unidentifiable subjective contribution, a creative-intuitive act of some kind.

In this sense, Poincaré can be placed in a broadly conceived "continental" framework. His description of mathematical invention involves, among other things, a sense of harmony and elegance, an esthetical sense expressed in the metaphorical superstructure of "imagery" used in creative reasoning. In a way, he argues that the process of "formal" reasoning cannot be fully separated from an emotive reaction. This is apparently of some interest to contemporary neuroscience. For example, American neurologist Antonio Damasio's 1994 book *Descartes' Error: Emotion, Reason, and the Human Brain* includes a brief discussion of Poincaré and notes that research bears out the idea of interconnectedness of emotive and formal thought processes.[10]

Despite his romantic notions about the process of scientific invention, Poincaré has contributed to science in ways that are relevant to physics and mathematics even today, almost a full century after his death. He was a pioneer of what we now call chaos theory; his critique of absolute time and of the notion of simultaneity, and his model of non-Euclidean geometry, anticipated Einstein's work on relativity theory; he seems to have been the first to suggest the speed of light as a meaningful limit to physical theories. It has even been said that Poincaré was one of the two most influential mathematicians of the twentieth century. Nevertheless, his reservations toward what would later become a widespread (mathematical) ideology were met by zealous resistance.

Let us look at another part of Poincaré's critique, the part in which we could find the early bud of some "postmodern" methods. The basic issues are again continuity and identity. Once more, continuity appears as something that is not logically grounded. Poincaré goes even further and observes that the same can be said of the concept of identity itself. His objective in doing so is to demonstrate the untenability of the notion that logic (or any other linguistic edifice) is the sole basis of mathematics, and to support his idea that continuity and identity are intuitively motivated.

Poincaré speaks of "the intuition of the group of continuous displacements of objects in space." The technical notion of a group is of little importance here, so the word can be safely ignored. What matters in the present context is that Poincaré argues that the notion of identity is construed not directly, but "negatively," as *invariance*, as remaining *un*changed with respect to possible transformations. (As far as mathematical content of this approach is concerned, Poincaré does not stand alone. For instance, the German mathematician Felix Klein proposed in 1872 to consider geometry as precisely the study of objects that are invariant under some group of transformations.)

Here is what Poincaré says in the article "On the foundations of geometry" (1898):

> What we call geometry is nothing but the study of formal properties of a certain continuous group [. . .]. The notion of this continuous group exists in our mind prior to all experience; but the assertion is no less true of the notion of many other continuous groups; for example, that which corresponds to the geometry of Lobachevsky. There are, accordingly, several geometries possible, and it remains to be seen how a choice is made between them. Among the continuous mathematical groups which our mind can construct, we choose that which deviates the least from that rough group, analogous to the physical continuum, which experience has brought to our knowledge as the group of displacements. *Our choice is therefore not imposed by experience. It is simply guided by experience.* But it remains free; we choose this geometry, not because it is more true, but because it is the more convenient.[11]

It is perhaps because of statements such as the last one in the quote above that Poincaré is sometimes described as a "conventionalist," that is, someone who thinks that all science is based on convention chosen for its simplicity or practical value.

Poincaré's ideas, I think, cut much deeper than that. In an article from 1891, he notes that Euclidean axioms are implicit and faulty definitions. In accepting them, we are tacitly assuming that the geometry we are talking about is the geometry of *solid* bodies. For example, what does Euclid mean by saying that two figures are equal if they can be superimposed? Poincaré observes that in order to superimpose them, one of them has to be displaced until it coincides with the other. He continues: "But how must it be displaced? If we asked this question, no doubt we should be told that it ought to be done without deforming it, as an invariable solid is displaced."[12]

So the hidden assumption surfaces now: "Objects" are invariable bodies. Poincaré concludes that the axioms provide us with a faulty definition of objects, giving us a vicious circle that defines nothing. "Such axioms," he wrote, "would be utterly meaningless to a being living in a world in which there are only fluids."[13] Thus, the very concept of identity of geometric objects, which every logical formalization takes for granted, is not a priori given within that formalization. It is presupposed.[14]

Let me pause here and insert some revisionist remarks that could allow us to compare Poincaré's ideas with those of my chosen set of pre-postmodern "postmodernists."

First, it seems that there is something of a family resemblance between Poincaré's reasoning and Wittgenstein's indictment of the notion of a thing being identical with itself. Poincaré notes that identity is conceived as the property of remaining invariant while being displaced. This seems to create a problem with regard to the identity of a *single* isolated object. The object would have to be, so to speak, separated from itself in order to be superimposed over itself.

In roughly the same vein, in paragraph 216 of *Philosophical Investigations*, Wittgenstein wrote: "'A thing is identical with itself.'—There is no finer example of a useless proposition [. . .]. It is as if in imagination we put a thing into its own shape and saw that it fitted." The difference between the two arguments is merely ideological. Wittgeinstein's "deconstruction of self-identity" is based on the *pragmatic* observation that the proposition in question is *useless*.

Second, I would like to draw your attention to the fact that Poincaré's strategy at least to a certain extent anticipates Heidegger's idea of "hermeneutics." Heidegger's approach takes as its proclaimed task to uncover the *un*said, the implicit, the omitted, the hidden presuppositions of its object of study, to uncover precisely that which comes *before* every act of reasoning the existential experience on which reasoning is based. (The stuff, in other words, that is handed down to us through the "clearing of Being.")[15]

In the terminology of Heidegger's student Hans Georg Gadamer—who is not a postmodern theorist but who occasionally veers in the direction of "continental anti-rationalism"—this could be translated as the uncovering of the "prejudices" of a certain tradition. These prejudices, a historically relativized edition of Husserl's intuitive core of the life-world, are constitutive of a tradition and tend to be taken for granted. They form a body of assumptions or beliefs that are not part of reasoning itself but underlie all reasoning practices and are the unstated condition of their possibility. It is the simple hypotheses of which one must be most wary, to paraphrase Poincaré, because these are the ones that have the most chances of passing unnoticed. The question of these "hidden" geometrical prejudices is indeed of interest to science.

For instance, one of the fairly standard reference works on mathematical physics asks the following question about what the authors call *pregeometry*: "What line of thought could ever be imagined as *leading* to four dimensions—or to any dimensionality at all—out of more primitive considerations?"[16] This is the kind of question that likely would have delighted Husserl and Weyl and Poincaré, maybe even Heidegger and Brouwer.

Poincaré's discovery of geometric prejudices could be formulated as follows. The identity of an object of geometry is not given by the logical structure imposed by the axioms: It is only a tacit *assumption*; it is the unsaid, the "before," that comes prior to logic. If we look at geometry as a formal structure governed by some rules, then there is no *logical* justification for our assumption that these "objects" we think we are studying have a permanent meaning or identity. The axioms are not self-sufficient, because they are already "prejudiced" by a belief in the self-identity of objects they are supposed to govern. So there is *first* a prejudicial and possibly subconscious act of "will" or "representation" or "interpretation," and only *then* does logic begin to apply.

The idea was formulated nicely by Nietzsche: "As a matter of fact, logic (like geometry and arithmetic) only holds good of *assumed existences* which we have created." As a further comparison, consider Heidegger's article "Modern science, metaphysics, and mathematics." Heidegger wrote that the project of "the mathematical"—recall that the term denotes some *pre*condition of knowledge, a space in which we "always already move" but that itself cannot be fully formalized—"*first* opens up a domain where things show themselves." Then he says: "[I]n this projection is posited that which things are taken *as* [. . .]."[17]

We operate, as it were, with some mathematical prejudices in the background. Otherwise no thing could show itself *as* a thing. These prejudices can be "thematized," but they cannot be logically justified or fully formalized.

Poincaré maintained that the identity of objects can only be continually *motivated* by intuition, in the sense that the invariance of an object under some group of transformations is ceaselessly reestablished perceptually. Thus, the identity of objects is a continually motivated *hypothetical* judgment. It is "constructed" as an invariance under transformations and is related to the construction of continuity. The notion of the mathematical continuum, Poincaré wrote in an 1894 article, is "created entirely by the mind, but it is experiment that has provided the opportunity."[18]

Let us return to Poincaré's comments about identity and unpack them as follows. The very concept of identity of an object depends on the possibility of it being different. Whether the circle is "really" a circle or a stage in the history of a pulsating ellipse cannot be said until it is given a chance (flow of time, "history") to change. A familiar corollary follows: Identity can be established with absolute certainty only from beyond history. Hence, the notion of identity is at best *motivated*, but it is not *grounded*.

I already noted that this reflection, in a suitably poeticized form, seems to be the starting point of parts of postmodern theory: As long as we remain within the confines of history, difference is somehow primary with regard to identity; identity is "not present to itself" because it is established through difference, which conceptually precedes it; and so on. (These observations are frequently accepted quite uncritically by an assortment of Derrida's "followers," who managed to turn them into an emptily conceived fetish.)

Once again, as in the case of Brouwer and Weyl, the ideological background of postmodern variants of this type of argument is not quite the same as Poincaré's, although the techniques are not entirely dissimilar. Whereas intuitionist mathematicians utilized time and difference to indicate the insufficiency of formal (linguistic or logical) structures to account for the sense of continuity and of identity of objects—and thus argued for the necessity of intuition or at any rate of some extra-logical involvement of the individual—parts of postmodern theory simply accept the "paranormal" consequences of dissolving into the fog of identity and continuity. I will return to contemporary variations on this theme later. For now, let us stay with Poincaré.

Unlike Brouwer's individualism and Husserl's ahistorical primordial intuition, identity of spatial objects is for Poincaré not entirely the conquest of the individual. It is a "conquest of the human race" that we have the intuition of the group of continuous displacements. For Poincaré, this intuition is due to the body as much as it may be due to the mind. In a similar vein, Bergson wrote of the "instinct," Husserl wrote of the "living body" moving through the life-world, while the twentieth-century French philosopher Maurice Merleau-Ponty held that the body's knowledge of the world is older than that of the intellect. Poincaré explicitly places this preintellectual knowledge in the context of evolution, and thereby—like, say, Heidegger—*into* history.

When we study geometry, says Poincaré, we are studying (our) bodies as they are presented to us: "In reality, space is therefore amorphous, a flaccid form, without rigidity, which is adaptable to everything, it has no properties of its own. To geometrize is to study the properties of our instruments, that is, of solid bodies."[19]

Let us now consider another relevant issue that Poincaré (and subsequently Weyl) brought up: the general concern with what he thought was circular reasoning, "begging the question," *petitio principii*, not only in geometry. We already saw some of the sources of Poincaré's interest in this problem. Poincaré notes that objects of geometry are defined in a circular way. We are using axioms to pinpoint what these objects "are," yet we are still making assumptions about these objects in the very notion of identity that is implicit in those axioms.

But perhaps it is possible to argue that identity of an object is indeed not "present" to the object itself. Perhaps its identity is granted through its relationship with all the other objects from which it is different. In this case, one might have to give up the idea of the absolute primacy of self-identity, but identity could at least be structurally deduced from "difference," that is, from a thing's relationship with *other* things within some well-regulated symbolic structure, such as a mathematical system. Such a view is usually associated with the work of the Swiss linguist Ferdinand de Saussure.

It may have been this question, among other related considerations, that led Poincaré to single out what is now known as "impredicative definitions."

These would be the definitions whereby an object is defined by invoking a reference to the totality of objects to which it belongs. For example, an impredicative d definition would be to define 1 by stating that 1 is the positive integer such that 1+2=3, 1+3=4, 1+4=5, and so on. The "trouble" with such definitions is that my knowledge of the defined thing depends on my knowledge of the totality to which it belongs. This is open to some objections, at least when dealing with totalities that have unboundedly many elements, which mathematicians frequently encounter. I cannot know the totality of integers, so if my definition of 1 involves all integers, then I cannot honestly say that I know 1.

Worse still, the state of my knowledge of what 1 is—say, at the stage when I have seen a billion integers—seems to be "mutable." Pure logic does not guarantee that my understanding of 1 shall not change in the future, when I have encountered two billion integers. The problem is especially acute for structures that are in some sense "generative," for example, the generative grammar of a language, which produces new elements over time. Then the impredicatively defined identity of the structural units may well have to be readjusted with the introduction of new units, and it cannot be a priori guaranteed that this operation will preserve their earlier identity.

An example that comes to mind is the value of money. If the state started printing money beyond reason, the value of a monetary unit would dwindle faster than a shooting star. Money derives its value not from the piece of paper on which it is printed, but, simplifying in the extreme, from its relationship with other such units. Few people have an idea of the total monetary mass currently in circulation, and most of us readily make the assumption that it is being kept under control. However, there *are* countries where unpleasant experiences with impredicatively defined monetary units are more than a mere theoretical possibility.

As a variation of this example, imagine a world where all or most things are both perfectly reproducible and impredicatively defined. The place could in theory be rife with inflationary phenomena of just about every sort. Concepts such as "implosion" might gain unexpected prominence. Endless reproduction of the simulated "real" could devalue it to such a degree that mystical principles like "virtuality is real" might begin to seem axiomatic.[20]

There is nothing *necessary* about any of this, but it seems to be a general feature of impredicative definitions that such things *could* happen. The identity of objects defined in such a manner is not guaranteed to remain immutable by the logical structure itself. Their identity is not present "in" them, but is granted by their structural relationships with other things. Therefore, introduction of new elements into the structure could very well make a difference.

The empirical fact that I think of such objects as having some identity and the empirical fact that I tend to regard my knowledge of them as having some continuity are facts that cannot be deduced logically. What is at work here, as in the case of our faith in money, is a principle of *induction* and not *deduction*.[21]

This inductive principle, like the notion of continuity and the identity of geometric objects, is for Poincaré the fundamental characteristic of human thinking and cannot be reduced to pure logic. *Without* the givenness of such a principle, things might fluctuate in a manner that cannot be a priori controlled—such as, say, the value of Yugoslav currency.

Thus, we are again facing the notion of an "open object," only this time this possibility presents itself not in the mysterious continuum, but in a discrete structure such as arithmetic or a "generative grammar." And so, if some such structure were required to provide for the identity of its units *itself*, the identity of those units may end up being open, their meaning unstable and "mutable." (A similar argument, applied to the formalistic excesses of continental structuralism, yields a result that might be called Derrida's basic theorem. See chapter 10.)

Poincaré's views on the problem of "mutability of meaning" are summed up in this paragraph by the American philosopher Warren Goldfarb:

> Since, for Poincaré, definability has no formal analysis, no prior limit can be put on the objects that may, one day, be subject to our reasoning. [. . .] A universal theorem does not relate to all objects, imaginable or not, in such a range; rather, it asserts only that each particular case of the theorem—each case defined in a finite number of words that will be considered by mathematicians or by succeeding generations of mathematicians—can be verified. [. . .] *Immutability is not secured by reference to a logical structure present from the start.*"[22]

Mathematical "text" is not *about* anything fixed. Precisely because of this mutability, mathematical meaning must always be constructed anew, by individual mathematicians and by succeeding generations of mathematicians. As an interesting comparison, let me note that Saussure—a thinker whose influence on continental philosophy I will address in detail later—expressed a rather similar opinion: "What has escaped philosophers and logicians is that from the moment a system of symbols is independent of the objects designated it is itself subject to undergoing displacements that are incalculable for the logician."[23]

To sum up, Poincaré brought up the following important problems: the ungroundedness of the very concept of identity of a single object, and its dependence on difference and intuition; the possibility that even in a well-regulated formal system the identity of an object cannot be guaranteed by its difference from all the other objects in the totality of that system; and the possibility that mathematical texts do not have a secure and immutable meaning, but are, on the contrary, always interpreted by individual mathematicians and succeeding generations of mathematics.

Let me conclude, here, this initial sketch of the intuitionist critiques that I think are relevant to understanding postmodern thought. I will return to these issues later to see whether the critiques of "logocentric" mathematics can be seen as the methodological precursors of some more contemporary lines of reasoning. Before that, we should look into the fate of an important counterproposal, Hilbert's program, its influence on continental thought, and the role it had in the transformation of mathematical intuition into an abstract computational device.

6
THE EXPIRED SUBJECT

The human being is eventually able to step aside and let a machine take his place.
—G.W.F. Hegel

It has been said that the German mathematician David Hilbert was one of the two most influential mathematicians of the twentieth century (along with Poincaré). The influence of his proposal for dealing with the question of mathematical foundations extends well beyond mathematics.

Considering Hilbert's program and continental variations on the theme will permit us to familiarize ourselves with the formalist sources of postmodern thought, which we will consider in more detail later. Along the way, we will glance at some side products of Hilbert's endeavor: the appearance of an abstract computational device as the model of human intelligence and the notions of incompleteness and randomness.

6.1. Empire of Signs

Hilbert, like Kant, comes from Königsberg. In a pleasing symmetry, Hilbert's first philosophically significant contribution appeared in the form of his 1899 book *Foundations of Geometry*. The most interesting feature of this work is the fact that Hilbert programmatically left the primitive concepts—point, line, circle—undefined. All that matters, ultimately, are the interrelationships of these unspecified objects, seen as symbolic elements of an abstract structure regulated by axioms. Geometry was to be regarded as a fully formalized axiomatic theory, independent of the interpretation of its primitive symbols. Hilbert is reported to have said, presumably to emphasize this feature of his approach, that one must at all times be able to replace "points, lines, planes" by "tables, chairs, beer-mugs."

Let me formulate it in the following way: Only the formal structural relations among the "signifiers" are of interest. *What* these geometric signifiers signify is immaterial, because they could signify anything, for example, the concepts of chairs and beer-mugs. The relationship between the signifier and the signified is arbitrary.

Hilbert believed that all of mathematics should be formalized in this way. But in contrast to the earlier projects of Russell and Frege, Hilbert did not regard intuition as completely dispensable. He held on to the scheme proposed by Kant, although with such serious revisions that it would be somewhat hasty to classify Hilbert as a "Kantian" in any straightforward sense. Despite some reasonable doubts about Hilbert's devotion to Kant, a note of reserved approval of Kant's project appears in a paper from 1931:"Even if today we can no longer agree with Kant in the details, nevertheless the most general and fundamental idea of the Kantian epistemology retains its significance: to ascertain the intuitive a priori mode of thought [. . .]."[1]

Where Kant relied on the a priori intuitions of space and time, Hilbert invokes something like an a priori intuition of finitary structures, an intuition of signs. In a way it is analogous to Kantian a priori intuition, but it is cleansed of the unnecessary "anthropomorphic garbage." This is a famous apocryphal remark, which seems to have found an overzealous following among certain precursors of postmodern thought.

However, before we consider the ways in which Hilbert's statement could be taken to unfathomable extremes, we should glance at what it was that he wanted to keep of Kantian human-centered theory of knowledge. Let us look at the notion of "finitary intuition" first. In the article "The new grounding of mathematics" (1922), Hilbert wrote:

> [A]s a precondition for the application of logical inferences and for the activation of logical operations, something must already be given in representation: certain extralogical discrete objects, which exist intuitively as immediate experience prior to all thought. [. . .] The solid philosophical attitude that I think is required for the grounding of pure mathematics—as well as for all scientific thought, understanding and communication—is this: *In the beginning there was the sign.*"[2]

Let me digress here, briefly. It is not difficult to understand why there would not be much of a consensus between Hilbert's school and romanticists like Brouwer. Hilbert's statement that *"In the beginning there was the sign"*—italicized in the original text—quite openly challenges the romanticist notion that action in some sense precedes knowledge. Fichte, for example, wrote that "[w]e do not act because we know; we know because we are called upon to act."[3] Even more interestingly, Hilbert's assertion can be construed as an ironic play on a phrase from Goethe's *Faust*, "In the beginning there was the *act*" (which is itself a heretical rendition of the Gospel According to St. John, where it is stated that "In the beginning was the Word").

So even though Hilbert's "finitary intuition" could have been tolerable to intuitionism, some important ideological discrepancies were lurking in the background. This fissure grew larger and more prominent as the "war of factions" continued.

I mention this because Hilbert's ideas about intuition would eventually turn into something altogether different from the notions with which we are familiar from the preceding chapter. Brouwer's was an intuition of the continuum, Hilbert's was an intuition of discrete, finitary objects. These finitary objects could be regarded as intuitively graspable forms (shapes, *Gestalten*), symbolical "phenomena."

So far, we have followed Hilbert in his claim that these forms are immediately available to us. Whatever these discrete intuitively given objects are—their precise nature turns out to be something of a mystery—our judgments about them could in Kantian terms be called "genuine judgments." In contrast to "ideal judgments" of general logic, which deal with potentially empty concepts and serve only as guiding or regulative principles, genuine judgments are not empty and constitute actual knowledge.

For Hilbert, in what appears to be a structural analogy with Kant's theory of knowledge, mathematics is separated into *real* and *ideal* parts. Real mathematics consists of genuine judgments and evidence of which our knowledge is constituted: finite structures, finite sequences of symbols ("proof-figures"), and judgments about them. As for ideal mathematics, it serves to stimulate and guide the growth of knowledge but is not part of knowledge proper. Hilbert's view of ideal judgments thus parallels Kant's explanation from the *Critique of Pure Reason*:

> [Ideal judgments do not] prescribe any law for objects and do not contain any general ground of the possibility of knowing or of determining objects as such [. . .]. [They] are merely subjective laws for the orderly management of the possessions of our understanding."[4]

Even so, such ideal judgments, and ideal "objects" that are on occasion produced by the application of ideal judgments, are a necessary part of reasoning. Rather than prohibiting their use, Hilbert argues that we should refrain from assigning "objective" meaning to such statements. That is not to say that Hilbert aimed to discard the entire notion of meaning, at least not initially. But this ascetic requirement is in all likelihood the source of the characterization of Hilbert's approach as a formalism that regards mathematics as an empty formula game.

This description apparently derives from certain remarks made by Hilbert's "archnemesis," Brouwer, in an article from 1912. It was then taken over by Russell. Weyl, a former student of Hilbert's, objected to the reduction of mathematics to purely formal, symbolic constructions akin to an "arbitrary game in the void proposed by the more extreme branches of modern art."[5] Similar sentiments are expressed, though more discreetly, in Husserl's *The Crisis of European Sciences* (1936), where the approach is indicted as the loss of mathematical meaning through "mechanization" or "technization." Such formalism—Husserl gentlemanly avoids naming names—reduces mathematics to something not essentially different from a "game of cards or chess."[6] Poincaré, too, raised various objections.

When so many people protest, one wonders whether there may be something to it. It seems, in hindsight, that Hilbert's school *eventually* painted itself into a corner that its critics described as formalism. However, initially at least, ideal mathematics was for Hilbert not an arbitrary game any more than regulative ideas had been arbitrary games for Kant. They reflect the structure of human thought, which *of necessity* transgresses what is given to it in phenomenal experience. So, was Hilbert a "Kantian" or a "formalist"? The issue is still debated. Let us take a closer look at his proposal.

No one, not even Hilbert, can deny that mathematical practice is teeming with analogical reasoning, metaphorical attributions of meaning to symbols, and heuristics of every imaginable sort. It was for "philosophical" purposes that Hilbert thought mathematics should be viewed as a purely formal, uninterpreted system of symbols. This move avoids some nagging questions about our beliefs regarding the nature of mathematical objects. Hilbert wants to sidestep the entire controversy between intuitionists' melting-hot universe of open objects and Platonists' frozen world of immutable forms. He has no interest in choosing sides in a speculative-philosophical conflict.

A way of conveniently avoiding the issue is to look at mathematics as a symbolic language whose objects—except for the bare minimum of intuitively given "Gestalts"—are grammatical dummies. They are like the "it" in the statement "it is raining." It makes little sense to ask *what* it is that is raining.

Incidentally, here we can discern what seems to be the source of Wittgenstein's concept of a "language game." Analogous to Hilbert's "formula-games," a language game consists of moves or utterances that are in some cases "about nothing." For Wittgenstein, the moves are justified not by a single unifying meta-game, but by their utility in a given "local" context. But let us stay with Hilbert for the time being. Wittgenstein's work was in part built on the ruins of Hilbert's attempt to construct a global "game" that would unify and justify all other mathematical "games."[7] Let us see how Hilbert planned to do that.

We are now taking for granted that most mathematical objects are grammatical dummies. We have a formal grammar of ideal mathematics, a system of uninterpreted "ideal" elements governed by some rules. Deductions within this system are finite sequences of symbols. They are graspable as finite Gestalts and are therefore among those precious rare "real" objects. As such, they are subject to *genuine* judgments that are part of the "real," intuitively given mathematics. We hope we can establish that none of these formal deductions—considered *themselves* as intuitively given objects, even though they occur in a "meaningless" formal game—ends up conflicting with any of the others.

So Hilbert's plan was to develop a mathematics *of* mathematics, called meta-mathematics or proof theory. Its task would be to establish that no deduction within the formal system of ideal mathematics leads to a contradiction. It would be the referee of all mathematical language games.

Meta-mathematics would thus serve to guarantee three important things: first, that the system that encodes the formal structure of abstract ("ideal") mathematical reasoning is consistent, that it is a compatible, noncontradictory extension of genuine reasoning. Second, it should guarantee, and this is directly contrary to Brouwer's claims, that at least a part of mathematical *communication* is undoubtedly certain: The finitist statements of genuine mathematics deal with objects that are accessible equally and universally to everyone in the (mathematical) community. Finally, if the first two points could be established, the program would render the *philosophical* question about the meaning of mathematical symbolism considerably less acute, if not downright irrelevant (at least from Hilbert's point of view). If challenged to legitimize itself rationally, mathematics could answer that it is at least free of contradictions, and that it therefore investigates various "logically possible worlds."

It looks like a good plan, but there is something strangely ambiguous about it. On the one hand, Hilbert associates himself with a species of Kantism and seems willing to enter explicitly philosophical debates about mathematics. On the other hand, when pressed to give some details about the mysterious finitary intuition, he apparently assumes the role of a practical scientist who is only interested in securing the consistency of mathematics but not in philosophical speculation. It is as if the philosophical question of the meaning of mathematics was for Hilbert simply one more item on a list of mathematical problems to be solved. (Hilbert did indeed believe that every mathematical problem can and will be solved. "We *will* know," he says, "we *must* know." From this point of view, it was a matter of putting the question about rational legitimation of mathematics into a *mathematical* form. It would be solved some day.)

I think that it was the second viewpoint that eventually took over. Despite frequent references to Kant and numerous philosophical discussions, the prevailing concern of Hilbert's program was to justify the correctness of the mathematical method *itself* as it occurs in accepted scientific practice. However it may have started out, finitary intuition ultimately became not the source of "truth" or "meaning" but of *social consensus*: It is the minimum that no practitioner of mathematics can reasonably deny.

Once this "social contract" is in place, the indubitable meta-mathematics would be used to establish that ideal mathematics is *consistent*, in the sense of not contradicting itself, and *complete*, in the sense of being capable ("in principle") of proving or refuting any given mathematical statement. Then the vague notions of truth and meaning could be eliminated and replaced by the concept of formal demonstrability. The "truth," or some formal surrogate of it, is entirely *in* the method.

It is here that we should seek the sources of at least some of the objections raised by several philosophers from Hilbert's spatiotemporal neighborhood. Even if his program were a complete success—which it ultimately was not—it would establish *only* that we can manipulate certain finitary structures without producing a contradiction. In this case Hilbert would have at best managed to legitimize Leibniz's "blind thought," the sort of thing that is now called symbolic manipulation. Leibniz himself attempted such a legitimization. The preestablished harmony, that is, the ultimate consistency of the universe, was guaranteed by the existence of God—which Leibniz famously proved. (Recall from chapter 2 that Kant later criticized Leibniz's existential proof: "The celebrated Leibniz," said Kant, not without a tinge of irony, "is far from having succeeded in what he plumed himself on achieving.")

Even though Hilbert's would have been a truly impressive achievement had it been successful, it would still leave some important questions unanswered. First of all, why does mathematics *work*? Second, if meta-mathematics investigates *only* logical consistency of purely formal structures, would that not amount to a complete disregard for the specifically human and historical aspects of the development of mathematical systems, severing them from of all motivation, thus rendering them meaningless? This, roughly, is Husserl's objection from *The Crisis of European Sciences*.

Analogous questions were raised by the neo-Kantian philosopher Leonard Nelson, who—through an effort of Hilbert himself—inherited Husserl's chair in Göttingen when the latter left for Freiburg. Hilbert's school and Nelson's group, of which Hilbert was an associate member, ended up in an interesting exchange over the role of meta-mathematics and the nature of Hilbert's "finitary intuition." Ultimately the two groups agreed to disagree. Nelson argued that even the intuition of finitary structures is in fact an intuition of space, because symbols, whatever they are, must finally be written out and apprehended as spatial entities.

Having a spatial intuition, even a minimal version of it, would permit at least *some* form of intuitive motivation of postulates. But Hilbert's school explicitly rejected the first part of this proposal: Logical consistency is all that matters, period. Nelson eventually expressed discontent with Hilbert school's "nihilism."[8]

What could be the motivation for such an uncompromising position of the Hilbert school? It seems to be this: The primary task of a truly scientific philosophy should be the rigorous study of methodologies *themselves*, not the justification of *our* knowledge obtained through these methodologies. Such considerations would involve us in unscientific speculations about human abilities and motivations, which are basically none of "our" business. At the very least, we should forget the romantic nonsense about creativity and other funny activities of individuals who hold theories of knowledge hostage to the capacities of individual human beings.

I may have exaggerated a little in anticipation of our discussion of certain postmodern ideas, and some people might protest my connecting such "anti-humanist" notions with Hilbert's formalism. The word "formalism" thus acquires an ideological connotation that may have little to do with Hilbert ironic dismissal of "anthropocentric" philosophy. Nonetheless, some subsequent and infinitely more extreme attacks on all things anthropocentric seem partly to be rooted in Hilbert's school of thought. This possibility was noted already by Weyl. "If Hilbert's view prevails," he wrote in 1928, "then I see in this a decisive defeat for the philosophical attitude of pure phenomenology."[9] Things are not so gloomy for phenomenology, generally speaking. But Hilbert managed to banish it from mathematics, through no fault of his own: It appears that subjectivity and formalism are difficult to reconcile.

In any case, it does not matter so much what *Hilbert* meant or wanted, because it seems that his formalism was radicalized by Jean Cavaillès, a philosopher of science whose work was a formative influence on the continental structuralist tradition and thus on its postmodern offspring. We will look at Cavaillès's influential variation on Hilbert in more detail later. Following that lead, I will argue that Michel Foucault, similarly but more radically, proposes a thorough cleansing of the entire Western philosophy of its anthropocentric biases—in a somewhat extravagant application of the formalist idea—for reasons that are not so different from Hilbert's.

But let us stay with Hilbert for a while longer. We have yet to see how and why his program came to be regarded as a failure. Let us consider some important questions regarding Hilbert's approach.

The most pressing question is to determine the precise nature of finitary intuition. Unfortunately, Hilbert was somewhat vague about it. So vague, in fact, that the question is still being discussed without any hints of a decisive answer as to what he meant. But at least one thing is crystal clear: Whatever his ideas about intuition may have been, Hilbert definitely did not envision meta-mathematics as a "speculative" philosophy. He saw it as a kind of *mathe-*

matics. And so, whatever it is that finitary intuition gives us, it should be formulated in a rigorous mathematical manner.

This seems problematic. Would that not make meta-mathematics *part* of the grammar of "ideal" mathematics, which Hilbert regarded as consisting of symbols to which no meaning should be attached? Would a *logical* formulation of the intuition which he said was *prior* to the activation of logical inferences not amount to pulling the carpet from under our own feet? Maybe not. But we would be left with the kind of formalism that Brouwer, Weyl, Husserl, and even Russell complained about.

Let me quote the Austrian logician Kurt Gödel, whose celebrated incompleteness theorems of 1931 are usually regarded as having in some sense "destroyed" Hilbert's project: "How indeed could one think of *expressing* meta-mathematics *in* the mathematical systems themselves, if the latter are considered to consist of meaningless symbols which acquire some substitute of meaning *through* meta-mathematics [. . .]."[10]

This meshing of formal and intuitive levels of reasoning could make the entire mathematical enterprise vulnerable to some classical paradoxes. For instance, if the intuitive, meta-mathematical notion of truth could be fully formalized, then one might expect difficulties with statements such as "this statement is false." That is exactly the root canal that Gödel explored. He showed that if the notion of truth *were* definable mathematically, then one could import a version of the Liar paradox into mathematics. We would be able to formulate mathematically the self-referential assertion "this statement is false," and mathematics would be thus inconsistent.

Carrying out all of this involves an extremely pedantic numerical encoding of every sentence expressible in a formalized grammar. This is both difficult and tedious. I will illustrate it by turning to one of Gödel's favorite philosophers, Leibniz, who came up with the basic idea in the first place.

Leibniz wanted to construct a symbolic language that would, as he wrote, "make argument and calculation the same thing."[11] For this purpose, he assigned numbers to what might be called primitive or atomic concepts. These are the famous "characteristic numbers." There is a problem here, in that he would have to know all these primitive concepts ahead of time in order to enumerate them. You may recall from chapter 3 that Leibniz said of these "marvelous characteristic numbers" that he only *pretended* that they are already given. Well, he had to pretend. The universal encyclopedia of all concepts is not yet available, although Leibniz dreamed of editing one of those.

Unlike Leibniz, Gödel does not have to pretend. He is dealing with a rigorously formalized grammar, so he can reasonably try to enumerate all of its sentences. Since I am doing my best to avoid technicalities here, let us pretend, like Leibniz—let us take some concepts and assign numbers to them. The concept "rational" will be assigned the number 2. "Animal" would be assigned the number 3. "Man," being a rational animal, would be a 6, which is 2 times 3. A monkey would be an animal, but not a rational one, so its number should be anything divisible by 3 but not divisible by 2. Let us give it the number 9. I will stop here, but I hope you have some sense of the general idea.

Statements expressible in our "grammar" can be numerically encoded in this manner, and the intuitive truths on which they are based can thus be made to correspond to what is demonstrable in mathematics. Monkey is not a man, and, sure enough, the monkey-number (9) is not an integer multiple of the man-number (6). Monkey is an animal, and, naturally, the monkey-number (9) is an integer multiple of the animal-number (3). This is trivial. But if we now pretend that we have before us an encyclopedia of all primitive, atomic concepts, and an appropriate coding of this kind, then the truth of *every* sentence of our language can be decided by algebraic manipulation or merely formal computation, that is, by "blind thought." That was Leibniz's dream. It came to an ironic climax in Gödel's proof, where it was actually carried out and utilized to show that it could not work the way Leibniz may have hoped.

Let us leave Leibniz aside now and look at Gödel's conclusions. So far, I seem to have established that the truth of sentences expressible in our mathematical grammar can be decided by computation. Gödel now observes that if our notion of what it means to be computable is sufficiently broad, then these computations can be made to talk about themselves. But computations are part of mathematics. In particular, it is now possible to establish the truth or falsity of the statement "this statement is false" in terms of computations, that is, inside the formal grammar of mathematics.

Once this is done, we have a paradox on our hands. So there are two choices: Either mathematics is contradictory, or the notion of truth is not definable in mathematics. The second option is clearly preferred. Truth is not definable formally. That is not a problem for Hilbert, because he wanted to replace this vague idea of "truth" by purely formal demonstrability. But it *would* be nice to make sure that statements we can formally demonstrate are not intuitively false. In other words, we do not want to be in the situation where our formal surrogate of truth contradicts the intuitive truth that it is supposed to replace. Perhaps we can ensure at least that much.

Gödel now points out that, unlike the notion of truth, the concept of formal demonstrability *is* definable in mathematics. A "proof" is a finite sequence of symbols of a certain kind. Therefore, demonstrability and truth are definitely not the same: One can be formally defined; the other cannot. So if we insist that *all* provable statements *are* true, then the converse cannot be true. Otherwise, truth and provability would be the same, which they are not. It follows that there are true mathematical statements that are not provable (in our formalization).

The argument in the preceding paragraph can be summed up, not without a touch of melodrama, as follows: Either mathematics is false, or there are true mathematical statements that are not provable (in a chosen formalization). This is usually referred to as "incompleteness."

I am stating it in this way, without avoiding the notion of intuitive truth, because that is how Gödel himself subsequently explained the heuristics behind his proof. It is also possible to formulate this result by invoking only formal demonstrability. In that edition, these theorems (there are two of them) basically state that Hilbert's program self-destructs *if* we choose to idealize the finitary intuition as a *sufficiently powerful* abstract-computational device.

This is just how it came to be idealized—I examine this important development in the next section—and so Hilbert's program came to be regarded as a failure. Gödel was extremely cautious about drawing this conclusion, and initially resisted it. So this "failure" is not absolute, and it is still a matter of discussion. It is contingent upon a particular formalization of intuition, a formalization that reduces it to some sort of a computer.

Gödel's heuristic explanation of his theorems involves the notion of objective mathematical truth, and even an intuition that comes along with it. That is a decidedly antiformalist idea, which he kept to himself until after Hilbert died. From some remarks he made subsequently, one can surmise that Gödel saw the root of "incompleteness" in the fact that the conceptualization of mathematical truth goes beyond what is accessible to us in a particular formal language. In fact, Gödel says that many formal languages face precisely the same difficulty: "[T]he concept of truth of sentences of [a language] cannot be defined in [that language]."[12] This is a theorem of the Polish logician Alfred Tarski, proved independently and published in 1933.

What is so remarkable about this is that Gödel—instead of discarding objective truth and intuitive insight, as the formalist fashion commanded back in those days—notes completely calmly that mathematical truth *is* objective and even intuitively "available," but cannot be crammed into language. It would be nice to know more about Gödel's philosophical outlook. Unfortunately, here as elsewhere, he was unusually cautious and circumspect. It seems that Gödel defended a strongly Platonist understanding of mathematics, that is, regarded abstract mathematical objects as objectively existing things and mathematical theorems as expressing objective truths about them.[13]

Nonetheless, it is known with certainty that Gödel was a serious reader of Leibniz and Kant, and of Husserl, whom he apparently held in particularly high regard. I do not know whether Gödel's notion of objective mathematical truth has anything to do with Husserl's. Still, I mention Gödel's reading list because it turns out that his incompleteness theorems have been wielded by some influential continental philosophers as a weapon aimed at Husserl's and Kant's anthropocentrism. I will examine in chapter 7 how this initial assault on "anthropocentric" philosophy unfolded, and how it made its way into postmodern thought.

Let us first have a look at what became of finitary intuition and how these developments seem to have affected science.

6.2. Mechanical Bride

All philosophical reservations aside, finding a mathematical model of Hilbert's finitary intuition was a crucial question that Hilbert's program faced. Clearly, Hilbert's school could not do what it set out to do, which was to study formal theories by employing only *genuine*, finitary methods, if there were no rigorous mathematical formulation of finitary intuition. Relying on informal intuitive notions on the *meta*-mathematical level provides as little comfort as does their application on the mathematical level: It simply displaces the original question.

Consideration of this problem led several mathematicians working in the 1930s to look into what it is that constitutes the human capacity for computation, in the broad sense of manipulating finite strings of symbols, and how that capacity could be formalized. British mathematician Alan Turing was studying the way people perform calculations in order to model an abstract computing machine based on that. Gödel worked on his own model, while the American logicians Alonzo Church and Emil Post each proposed their own version of what it means for something to be effectively (or "genuinely") computable. These were the theoretical prototypes of what we now call computers.

Turing argued, convincingly, that anything that could in principle be calculated by a human calculator can likewise be calculated by one of his abstract machines. These machines are rather like our contemporary computers, except that they can take any amount of memory and any length of time to do the calculations. Gödel was for a while skeptical about these models of effective computation, but he was convinced by Turing's analysis. Some time in the 1930s, it was proved that various proposed descriptions of computability were equivalent to each other. This gave some empirical weight to the following:

> Church-Turing Thesis. *The notion of being effectively calculable (in the intuitive sense) should be identified with the kind of calculation that can be carried out by a Turing machine.*

The Church-Turing thesis is not a theorem. It is a reasonable hypothesis. It cannot be proved. Its refutation, on the other hand, would involve finding something that we know is computable but is not computable by a Turing machine. Since Turing's notion of a calculating machine is based on a painstaking analysis of how humans perform computations, such a refutation seems unlikely. Therefore, most people eventually agreed to accept the thesis.

What this did for science and the culture of machine fetishism became clear only later. It was the first computational, physically realizable model of some processes that go on in the human mind. Cognitive sciences and the artificial intelligence camp make much of Turing machines and the Church-Turing thesis, so let me emphasize what, precisely, we have thus far: an *agreement* as to what it means to be computable or effectively calculable. It means, roughly, to be programmable on a kind of an idealized computer, that is, one with unbounded memory and an unlimited lifetime available for its computations.

We now stand on the threshold of an amusing ideological reversal: What was initially a convenient abstract model of the human capacity for computation—agreed to by mathematicians because they needed a formal model of finitary intuitive computations—would become something of which the human mind is a particular instance. It is not that the computer is modeled on the mind. It is the other way round: The mind is a model of the computer.

One might be tempted to observe that the tendency to worship entities in whose image we are all made is too strong for our own good. But that would perhaps be too cynical. One could nevertheless note that the notion of effective computability might not exhaust what goes on in the human mind. A Turing machine can do all the computations I could ever do, and infinitely many more computations on top of that. I might be able to do some things (not computations, of course) that it cannot do.

This may seem like a sad attempt at refuting the Church-Turing thesis. Rest assured that I am not trying anything of the sort. I might be questioning the way some people tend to understand and formulate the thesis, but not the essence of the thesis itself. Its essence, the way I understand it, is that the notion of what is intuitively (and hence humanly) computable is *captured* in the notion of the Turing machine. It indeed seems to be the case that everything intuitively computable is computable by a Turing machine. This means that if something *cannot* be computed by a Turing machine, then I would certainly not be able to compute it. That much is clear. I am complaining about the converse: If something *can* be computed by a Turing machine, I may well be unable to do it. It could be completely beyond me. For instance, I have a bounded amount of memory and limited time at my disposal. I get bored, I daydream, and so on.

So Turing-machine computability is an idealization, a convenient and perhaps a groundbreaking one, but an idealization nevertheless. It is like Deep Thought, the supercomputer from Douglas Adams's *The Hitchhiker's Guide to Galaxy*, which finished its computations and printed out the ultimate answer "42" *so* far into the future that no one even remembered what the original question had been.

The Turing machine is not just an idealization but a *very* specific kind of idealization. This abstract computer was designed exclusively to capture what I can compute, not everything I do. Turing definitely did not base his model on, say, a detailed study of the mating rituals of human beings. Nor did he base his model on a detailed study of the mysterious goings-on in the minds of Russian symbolist poets. His study convincingly argues that the machine captures the computational ability of human beings—it positively *over*estimates this ability, raising it up to the technically convenient levels of mathematical abstraction—and it was accepted for no other reason than that.

Thus, it seems that there are some important things to be taken into account before we start pronouncing that the mind is a Turing machine. Anything can be turned into mathematics, modeled by a formal system. Even weather can be simulated and predicted a couple of days ahead. But that does not mean that the weather is a piece of software on God's personal computer.

However, lately, this is not the way one sees the Church-Turing thesis called upon. What was for mathematicians originally a consensus about what it is that could reasonably be regarded as a description of the vague notion of "intuitively computable" became the definition of the human mind itself. What was once a convenient description is now a regular prescription.

Let us note a few things that seem to me beyond doubt. The notion of formal computability arose from attempts to model Hilbert's finitary intuition. The machine replaced intuition completely, by means of a consensus known as the Church-Turing thesis. Finally, it was declared that this abstract device is no less than the universal blueprint of the mind, a universal spirit of which we are but physical realizations.

Before we consider the ways in which some postmodern thinkers are striving to contribute to this symbolic euthanasia of the human subject, let us look at two more results about the limits of computability. They will become relevant later, and this is a convenient point to introduce the concept of randomness as explained by the American mathematician Gregory Chaitin.

To begin with, note that every formal system, every formalized grammar of some language, can be encoded by a Turing machine. All "proofs" and all grammatical sentences appear as a result of certain basic operations on strings of symbols. The rules of grammar can be made into a program for our idealized computer. In the case of natural language, these rules are elegantly spelled out in Chomsky's "X-bar theory." It does not matter, for our purposes, what exact form it takes. Suffice it to say that it is related to formalized grammars (and hence to idealized computing machines).

It therefore seems natural to ask if there is a program that could decide, for any given string of symbols, whether it is a grammatical construction or not. This seems easy. For example, I could write a program that derives all grammatical sentences, one by one, and checks whether the given sentence

The Expired Subject 79

eventually shows up on the list. If the sentence is grammatical, it will finally appear. But what if it is not grammatical? I would have to wait forever.

So I have to do a little better than applying the "brute force" method, which lists all grammatical constructions and tries to find my sentence on this long list. But this idea already got half of the job done. If something is grammatical, the "brute force" method will tell me that it is. It is the other half that causes trouble. Perhaps I could somehow refine the same idea to deal with that.

In effect, I have to be able to decide which of my program's computations halt and which do not, before I let the "search all" program have a try. If I know that the program stops for the given input, then I know that it *will* find my sentence on the list of grammatical ones. If I know that the program does not stop for the given input, then my sentence *will not* be found on the list and hence is not grammatical. But we know that there is no program that can make such a decision, generally speaking. (It depends on how complicated the grammar is. There are some for which such a decision can be made, and some for which it cannot.)

Applied to a mathematical "grammar," this could be interpreted as follows. If I identify the truth of a theorem with its provability—"provable" sentences are "grammatically correct," are constructed from the axiom-sentences by means of rules—then I cannot *in general* decide whether or not a statement is true. Generally speaking, there are statements whose truth is "undecidable."

This is bad enough, but Chaitin has demonstrated that the situation can be made even worse. Not only are there statements whose provability cannot be effectively decided, but also there are statements of whose provability we cannot say more than we could say about the outcome of tossing a coin: heads, provable; tails, not provable. That is the best we can do. This will become relevant later, so I would like to illustrate the basic ideas behind the proof.

To formulate Chaitin's result, note that we do not need to look at many different Turing machines that do different computations. Rather, we can look at a universal Turing machine, which can simulate any other—just like there are PC simulators for a Mac. Then any Turing machine can be identified with a program for the universal one. It does not matter what programming language we use, so I let us assume, for simplicity, that we are dealing with "machine code," that is, a string of 0s and 1s. Input values can be made part of the program, and output values are also strings of bits (0s and 1s). Now everything is in bits: input, program, output. Let us not forget that it is impossible to decide whether programs halt or not.

Chaitin asks the following question: What is the probability that a "randomly" chosen program will stop? This probability is called Ω, the "halting probability," and represents the likelihood that the program written by tossing a coin a number of times will be one of the halting programs. Recall that a program is just a sequence of 0s and 1s. If I choose 0 to represent heads and 1 to represent tails, I can write a "random" program by tossing a coin a few times.

Since everything is already expressed in 0/1 notation, it is convenient to look at the halting probability in the same notation: All the digits of Ω's "decimal" expansion are either 0 or 1. Now let us see why this number is not computable, in the sense that no program could generate the sequence of its digits.

Suppose we know the probability that a 2-bit program will halt. This probability is the number of halting 2-bit programs divided by 4 (since 4 is the number of all possible 2-bit programs: 00, 01, 10, 11). Hence, we know how many 2-bit programs actually stop. Say, for example, that we know that exactly three of them halt. We can then run all four 2-bit programs simultaneously and simply wait. We know that exactly three of them will stop, so if we patiently wait it out we will eventually know *which* three stopped. The one that did not stop after the three halting ones had, never will.

Thus, we are able to decide which 2-bit programs halt and which do not. This argument can be applied to longer programs. It follows that if we knew the probability that n-bit programs halt, for all n, we could then solve the Halting Problem. It takes some work, but a more sophisticated version of this idea can be used to show that if the digits of the number Ω were computable then we would be able to solve the Halting Problem. However, Turing proved that this problem is unsolvable. Therefore, I cannot come to know Ω by formal reasoning of any sort, because all formal languages can be made into a "grammar" whose rules are programmable on the universal Turing machine.

The halting probability Ω encapsulates all the information that could ever be encapsulated in any formal theory. We now have a single number—defined in an apparently reasonable manner—which is *so* uncomputable that it goes beyond the deductive power of any past, present, or future formalized theory.

This is not as mysterious as it may sound. In a way, Chaitin (whose arguments come from the 1960s) formalizes Borel's ironic remark from 1927, already quoted above:

> One could define [a] number by saying that each of [the] successive digits [of its decimal expansion] is equal to 0 or 1 according to whether the answer to some question or other is affirmative or negative. Moreover, it would be possible to order all the questions that can be asked in the French language by sorting them [. . .] as is done in dictionaries. Only those questions for which the answer is *yes* or *no* would be retained. The mere knowledge of the number thus defined would give answers to all past, present, and future enigmas of science, history, and curiosity.

Chaitin uses a programming language instead of French, but the idea is roughly the same. This number, Ω, is beyond formal methodology. Through this connection with Borel, Chaitin's results, much like Turing's original approach, seem to be closely tied with the problem of the continuum. Gödel's theorems also arose from his attempt (and failure) to justify the classical theory of the continuum. Discrete simulations of the continuum go only so far. They have their advantages and their shortcomings. One can suppress the shortcomings somehow, but they make their way into other places, causing incompleteness, undecidability, and even randomness.

But what does all of this have to do with randomness? It turns out that Chaitin's number Ω, the halting probability, is in a certain sense random. There are a few different descriptions of the term "random." Most of them have to do with complexity of things in one way or another, and Chaitin's description has to do with complexity of programs for our abstract machine.

We should take a glance at how this works. Randomness and chaos seem to be the darlings of the more popular forms of postmodernism that I will consider toward the end of this book. In fact, as soon as the next chapter, we will have a chance to observe how randomness enters the picture. Let us first consider one of several mathematical ways of looking at the notion of randomness.

First, let us look again at strings of bits, that is, of 0s and 1s. Chaitin defines the complexity of a string as the length of the shortest program that has that particular string as its output. For instance, the string consisting of a billion 0s is more concisely defined as a program consisting of the statement "print one billion 0s" than by actually writing out a billion 0s. So its complexity is significantly less than a billion. The information contained in that string can be *compressed*. Its information content can be captured in a string that is substantially shorter that the string itself, so we cannot say that it is completely random.

According to this view, a string would be considered random if its description cannot be compressed in such a way: We cannot define a random string of length N by a program that has length substantially less than N. (Otherwise it has some structure, so it is not random.) In a sense, these are then the most complex of strings. They are "random" because they resist reduction to simpler things. They are, so to speak, formally irreducible. Thus, a random string might be said to simulate an event whose "cause" (the program that generates it) cannot be reduced to a more primary "cause" (a simpler program that generates it). A random event "just happens."

A simple computation shows that random strings are also the most ubiquitous ones. For instance, only about 1 in 16 strings of a given length can be compressed by 4 bits. Only about 1 in 500 strings of a given length can be compressed by 9 bits. The more compressible the string, the more "structure" it has. Strings with more structure are less frequent. Those with less structure are more frequent. Hence, "randomness" is really not a very exciting phenomenon. What is exciting for mathematicians is that it can somehow be described, *modeled*, and hence studied.

Chaitin's theorem states that Ω is in a way *as* random as anything can get. It is incompressible in the sense that no program can generate its digits beyond a certain point. Whatever program I choose, there will be a digit of Ω, somewhere down the line, about which the program will tell me nothing. All I can say is that the digit is 0 or 1. If I think it is 0, I have an exactly 50-50 chance of being right. That is the best I can do.

Now let us equate truth with demonstrability and see what happens. Demonstrability always depends on some formalization, a "grammar" of some kind. The rules of this grammar can be encoded by a program, P, for the universal Turing machine. If our grammar is broad enough to express some basic mathematical notions, then we can define the number Ω by a grammatical sentence. We would like to generate the truths encapsulated in our formal system, that is, the "grammatical" sentences. We can do that: Just let the program P list all the grammatical sentences. Since Ω is defined in our formal grammar, we can ask questions about its digits. Is the 164th digit of Ω 0, or is it 1?

But P is just a program, and we know that no program can compute all the digits of Ω. So it follows that whatever my formalized grammar happens to be, there *will* be a sentence like "the nth digit of Ω is 0," which not only is undecidable but also is *exactly* as true as "the coin I tossed will produce heads." It's random. And so it seems that all sufficiently strong systems of formal reasoning have some randomness inscribed into them.

However, all this does not entail that truth *is* random and that everything is arbitrary. I said, quite casually, "let us equate truth and demonstrability and see what happens." As it turns out, there is too much at stake to be so casual about it. Let us see how far things can go if we are not extremely cautious.

7
THE VANISHING AUTHOR

Only a foolish artist can claim that his work is entirely his own.
—J.W. Goethe

We will embark on a longer excursion now. I would like to consider a different way of understanding the formalist position, and thus try to connect it with some apparently unrelated and extreme branches of continental thought. In particular, I would like to look into the possibility that Hilbert's "formalist" approach, filtered and transformed through the work of the French philosopher of science Jean Cavaillès, may have carried over an important impulse to continental philosophy. This alternative pathway to the postmodern turnabout will eventually lead us to the early work of the French historian and philosopher Michel Foucault.

To begin with, consider the possibility of a "Hegelian interpretation" of Hilbert. That, approximately, is Cavaillès's starting point, and so it shall be ours as well. Hegel's operation on Kant proceeds on two levels. First, it undermines the very root of Kant's setup, namely, the dualism between theory and application, knowledge and reality, transcendental and empirical. Second, it brings history into the picture.

Now, if we think of Hilbert's plan in terms of its self-proclaimed proximity to Kant, then we can think of Cavaillès's philosophy of science as, so to speak, doing a little Hegel on Hilbert's "Kant." In this manner, one might get a reasonably nice variation on formalism, as I hope to illustrate. However, as is true of most things, it can in a few easy steps be twisted into something extremely strange. But first, let me introduce Cavaillès and indicate why I think he can be viewed as bridging the great divide between Hilbert's formalism and certain parts of postmodern theory.

Cavaillès was a philosopher of science and mathematics, a critic of Husserl and Kant, and a (twice) decorated hero of the French resistance. The role of his work in the changes that took place in the French philosophical scene after World War II—he was executed in 1944 by a Nazi firing squad—is perhaps unfairly neglected outside of France. I will concentrate on his influence on Foucault, but it extends further than that. Skipping over unnecessary details, let me simply say that Cavaillès's work helped remove the "spell" that the intuitionist Bergson and the existentialist Sartre cast on French philosophy.[1]

What is interesting from my broadly conceived mathematical point of view is that Cavaillès's outspokenly anti-intuitionist and anti-existentialist stance, his tireless emphases on methodology and structure of science itself as opposed to the philosophy of the (human) subject, were at least to a certain extent influenced by Hilbert's formalism. Consider, for instance, the titles of his books: *Axiomatic Method and Formalism* (1938), *The Transfinite and the Continuous* (1943), *On Logic and the Theory of Science* (1943).

Hilbert's plan was to develop a meta-mathematics, a mathematics of proofs themselves as finite sequences of symbols, so that the vague notion of truth would be replaced by that of formal demonstrability. Cavaillès, similarly, spoke of "science of science," blurred meta-mathematics into mathematics, and maintained that the truth is *in* the demonstration, *in* the method itself.

To be sure, there are enormous differences between Hilbert's and Cavaillès's views, as I will show in a moment. But Cavaillès is not untrue to Hilbert in some respects, and he was not an amateur in matters mathematical. Cavaillès got his doctorate in mathematics and formal logic in 1938. This was *after* Hilbert's program was declared a failure due to Gödel's incompleteness theorems and Turing's undecidability theorem, which is important.

While Hilbert attacked the romanticism of Brouwer and Weyl—"No: Brouwer is not, as Weyl believes, the revolution"—Cavaillès was more thorough. He went after the "philosophy of the subject" in general, especially Husserl's and Kant's ahistorical intuitions. Cavaillès's science of science appears to be blessed with a "Hegelian" slant, which is clear from his rejection of the pure/applied dualism, his critiques of the philosophy of individual consciousness, and his concern for change and movement that constitute the structure through which science manifests itself *to itself*. This is not unlike Hegel's spirit, which through a dialectical movement comes to know itself *as* that very movement.

Science cannot be reduced to the intentions of individual scientists, but is an entity in itself. Applying this to the particular case of mathematics, we get the following picture. A theorem is not true because *someone* got an idea and then applied the universal, immutable laws of logic or mathematics, thereby proving the theorem. Rather, the "truth" of the theorem is in its very demonstration, which represents a necessary movement within the structure of science itself. "The true meaning of a theory is not in what is understood by the scientist," wrote Cavaillès in *On Logic and the Theory of Science*, "but in a conceptual becoming that cannot be halted."[2]

Following this idea, we come across Cavaillès's line that scientific progress is not a history of accumulation of truths but a perpetual revision through deepening and erasure. On this view, the task of historians of science is to study the *constitution* of truth as a historical concept within an era, rather than to study what was believed to *be* true in that era. Foucault says something similar in *The Archaeology of Knowledge* (1969): "[T]he knowledge of psychiatry in the nineteenth century is not the sum of what was thought to be true, but the whole set of practices, singularities, and deviations of which one could speak in psychiatric discourse."[3]

This similarity is more than a random coincidence. There is a strong historical link between Cavaillès and Foucault. Foucault himself acknowledges his debt to the French historian of science Georges Canguilhem, his mentor. Canguilhem, on the other hand, admired Cavaillès's work and personal courage—even wrote a book about him, *Vie et mort de Jean Cavaillès*—and was one of the people whose influence paved the way for postmodern theory in the somewhat rigid world of the Parisian academia. (One tends to forget that before it was successfully exported to the United States, most of what people now call "poststructuralism" and "postmodernism" had been a fairly obscure continental phenomenon.)

Now that I have at least indicated some traces of Cavaillès on Foucault's intellectual horizon, we should check whether Cavaillès's outlook can be seen as having grown out of Hilbert's. Let us start with some easy observations and follow their transformations and variations, keeping in mind that we are looking to find a place for Hilbert's formalism in the genesis of postmodern thought.

Hilbert was partly inspired by the method of "ideal elements," fictional entities that mathematicians add to their theories for some strange reason. I have in mind "imaginary" objects like, say, the square root of -1. The formalist idea is that these things are really nothing outside of the context of formulas in which they occur. They are grammatical dummies, produced by grammar itself.

The view that "the object is the product of the method," coupled with a distrust toward the individual anthropomorphic creatures who apply the method, is lurking at the background of formalism. Cavaillès simply radicalized a built-in feature of it.

Let us continue with the method of ideal elements. The minimum one must grant is that these symbolic constructs do not have any meaning independent of their formal-linguistic context. They are not names of ideas, nor do they have the power to summon up ideas when taken in isolation. For example, the letter i, traditionally reserved for the "imaginary unit," that is, the square root of -1, fails to evoke anything mathematical in the context of the following statement: *This is not a pipe*. That much is completely obvious, but let us look further.

A symbolic "object" such as the square root of -1 was originally not something that represented a ready-made idea. There was no idea present to begin with, so it could not be *re*presented. This object made its appearance in the process of finding a formula for the solutions of certain kinds of equations. Today, we would simply say that the square root of -1 *is* a solution. However, the sixteenth-century Italian algebraists who were looking to solve these equations—Girolamo Cardano, Niccolo Tartaglia, Scipione del Ferro, Rafael Bombelli—had no concept of square roots of negative numbers as objects. They applied the method *purely formally*, and these pseudo-objects started showing up. The entities that appeared in this way were considered a deviation of some kind. They were not "real" objects. (Cardano spoke even of negative integers as "fictional numbers," *numeri ficti*.)

These fictions were not what the method was *about*. Today it seems that we are talking about something when we invoke the square root of -1. But that was not always so. Certainly not in the sixteenth century. In fact, it was only over the course of a couple of centuries that the so-called "imaginary numbers" (the name is telling) became acceptable *as* objects of mathematical practices.

Thus, these previously unreal entities became acceptable objects of study. They became "real," or, to put it differently, the idea became sufficiently concrete through practice, over time. However, it is not the case that the *concept* of "imaginary numbers" is due to, say, Cardano or Tartaglia. They had no *idea* about it. But the notion is not due to Leibniz either—although he performed some "blind" algebraic operations with imaginary numbers—because it was in *some* sense "discovered" by those sixteenth-century Italian algebraists. So who should get the credit for the invention? (It is a little like asking Darwin to pinpoint the exact ape who first started walking upright.)

One way to resolve this copyright issue is to give the credit to history, to the mathematical community, or to the spirit of mathematics. This is far, but still not *too* far, from Hilbert, despite the Hegelian formulation. Hilbert was more of a communitarian than an individualist: "*We* will know, *we* must know," some day. But this "we will know" eventually turned into something like "an idealized computer will compute it," perhaps centuries from now. Such computations are not necessarily within the reach of a human being. Most of us have a relatively short shelf life. In this sense, a "Hegelian" interpretation seems closer to the formalist practice of replacing intuition by an idealized machine, despite the Hilbert school's incantations of (the thoroughly anthropocentric) Kant.

Cavaillès explicitly chose such an approach. The credit goes to science *itself*. In other words, there is no one in particular to whom we have to thank for imaginary numbers, the telephone, the computer, nuclear bombs, and so on. It was a natural, historical movement of science itself. (Foucault radicalizes this idea and argues that no text has an author.)

I am inclined to quote the physicist Niels Bohr here: "That's crazy, but not crazy enough to be right." Yet the approach is not entirely absurd. It explains the invention of, say, imaginary numbers, much better than the colloquial "a light bulb went off in someone's head" approach. So we must not dismiss it on instinct alone.

Let us look at the case of imaginary numbers again. One could argue that the very *possibility* of the formation of the idea of imaginary numbers was opened up in the process of applying a certain procedure for solving equations. Some Italians applied the method blindly, and *voila*. In this sense, the method historically precedes the formation of a socially accepted object.

I think this is undeniable. But there is a dangerous bit of formalistic ideology lurking just around the corner: denying human beings any creative role in this process. Imaginary numbers would have been discovered anyway, and it makes no difference by whom. The object is the product of the method, for which we are no more than narcissistic conduits. Removing the anthropomorphic rubbish from the narrative now seems almost natural. The credit should and does go to science itself.

If we take Cavaillès literally—and there are people, such as Foucault, who did just that—it seems that mathematical language gets extended of *its own* methodological necessity. *It* introduces completely formal idealizations, so the universe of mathematical objects is always in the process of formation.

The next step on our slippery slope toward postmodern formalism is to show that the same happens with the sacred concept of truth. To avoid unnecessary misunderstandings, let me say right now that the statement "all truth changes all the time" cannot be proved, period. Nonetheless, a "proof" of this statement can be simulated in the formalist setup sufficiently well to convince a number of people. Formalism equates truth with a linguistic surrogate called demonstrability, and language, as we know, changes. To support this obvious remark formally and make it into a statement about "truth," some form of the incompleteness theorem is introduced into the evidence.

We could invoke Tarski's theorem here. It states that no decent formal language can formulate its own notion of truth. Thus, language *itself* requires its extension to accommodate this deficiency. It requires ever higher concepts of truth. Hence, the process of formation of mathematical truth takes up the whole of history. So mathematical truth is utterly beyond individual human beings, who are finite. Kant and Husserl were wrong. The case is (supposedly) closed.

Individual consciousness cannot guarantee that it a priori knows mathematics. No individual, with the possible exception of Hegel and Francis Fukuyama, can even imagine the conceptual changes that extend infinitely through history. That is why Cavaillès wrote that the "true meaning of a theory is not in what is understood by the [individual] scientist, but in a conceptual becoming that cannot be halted." Meaning is in the method; method extends through all of history in an endless dialectical overcoming. (The spirit of Hegel, the old patriarch, seems to be laughing at us from the great beyond.)

Gödel expressed a relatively similar opinion in a footnote to his famous 1931 paper about the incompleteness of formal systems, which is likely Cavaillès's source of inspiration: "[T]he true reason for the incompleteness inherent in all formal systems of mathematics is that the formation of ever higher concepts can be continued into the transfinite."[4] So he, too, grants that there is a "conceptual becoming that cannot be halted." However, Gödel also wrote that these higher concepts are always judged in practice. That can only mean *human* practice—practice in a life-world where people speculate about higher concepts based on their intuitive sense of truth, simplicity, beauty, analogy, utility, experimental confirmation, fashion, cultural background, political or personal agendas, and, well, you name it. We have to do it the hard way, because we cannot prove what Husserl dreamed of proving, namely, that there is *a* part of our life-world that remains invariant throughout history.

We already know that much from Derrida's critique of Husserl, discussed in chapter 4. But now we are facing another argument, and it seems even stronger. Let me sum up in two sentences what Cavaillès seems to be claiming, or what his followers think he is claiming: The meaning of mathematics is *in* the endless historical process of its changes. Since mathematical truth *always changes*, it remains beyond the reach of individuals.

The above "proof," despite its references to the incompleteness theorems, claims more than it can deliver. It apparently claims that there is *in fact* no immutable core of mathematical truth—it always necessarily changes throughout all of history.

No one can prove such a thing. It makes sense only under the formalistic assumptions that truth is identical with provability and that meaning is in the method *itself*. Everything is always and only a matter of language. But this assumption is not beyond doubt. We know, for instance, that Gödel did not choose to go down that path. It seems to me that he had good reasons. Clearly, as Derrida told Husserl, you cannot prove that something (e.g., truth or intuition of time) is invariant in time, unless you observe it throughout *all* of time. Similarly, you cannot prove that *every*thing always necessarily changes, unless you know *every*thing.

Simply said, Cavaillès's argument fails to make good on its promise. So there is at least a glimmer of hope—Gödel, Husserl, Brouwer, Weyl, and Poincaré all expressed this hope in different ways—that mathematical practices are based on some (inner or outer) truth that remains invariant even though our formalization of it may indeed change. *If* there is such a thing at all, then it is beyond formal methodology.

Hence, the problem of how best to formalize the elusive truth, in which most mathematicians believe, is in essence a social issue. It involves politics, personal agendas, fashion, prestige and power of certain institutions, and other human affairs. Sometimes a "war of factions" occurs, as in the case of intuitionism. Sometimes a higher concept such as the Church-Turing thesis is introduced by a very few "wise men," who have a little talk, make a decision, and then let it trickle down. Perhaps not everyone accepts it. Those who do, do so for a variety of reasons. Some people accept it because it makes no difference to them (the "whatever" demographic). Others accept it because it *is* reasonable. Yet others, such as the crusaders of artificial intelligence, turn it into a religion. This process, as are most social interactions, is quite complex.

Foucault says this very nicely in *The Archaeology of Knowledge*: "In short, a proposition must fulfill some onerous and complex conditions before it can be admitted within a discipline; before it can be pronounced true or false, it must be, as Monsieur Canguilhem might say, 'within the true.'"[5]

I do not think that anyone denies this, but we have yet to see where Foucault is going. For now, we must grant this much: On the formalist assumption that the truth is in the method itself—and not at all in the people who apply the method—it becomes pointless to speak about "knowledge" and "truth." We cannot guarantee that the terms "knowledge" and "truth" retain sufficient continuity of meaning throughout history. Cavaillès is supposed to have demonstrated this in the case of mathematics: Mathematical truth (supposedly) always incessantly changes in history and thus remains beyond individual understanding. Foucault simply accepts this as a given, and he argues that the same remark applies to all sciences, humanities, and philosophies.

Foucault wants to draw some extremely radical conclusions based on his faith in Cavaillès's formalistic notions about the spirit of science. His proposals are slightly exorbitant and require a critical examination. (I would like to note that my criticisms are not "ad hominem," and that I find Foucault's work stimulating and interesting in its own way. I cannot do it justice here, and I will concentrate only on a few specific aspects of his oeuvre.)

As a point of departure on our journey toward Foucault's proposals, consider the effects of Cavaillès's viewpoint on a familiar example. Strictly speaking, what the sixteenth-century Italians did with imaginary numbers cannot be called knowledge or truth. It was not considered as such even by them, because they viewed it as they would view Russell and Whitehead's *Principia Mathematica*: a cryptic scripture that vaguely resembles mathematics. The formulas the sixteenth-century algebraists came up with are of course true for us, now, but that might be a retrospective illusion.

So if we want to have a proper *science* of science, if we want to do history the way positivism did sociology—and this is exactly what Foucault, supposedly ironically, declares as his objective: the return to "felicitous positivism"—then we cannot say that the dissident practices of Cardano and others *were* and *are* knowledge and truth. They just wrote down some baffling formulas involving objects that were not even considered to be objects at that time.

But if it was not knowledge or truth, then what was it? It was, as Foucault says, a discursive practice. This, I think, may be what Foucault was trying to convey when he wrote in *The Archaeology of Knowledge* that "[k]nowledge is that of which one can speak in a discursive practice, and which is specified by that fact: the domain constituted by the different objects that will or will not acquire a scientific status [. . .]."[6] Let us accept this terminological convention—that, basically, is all that it is. But since he introduced this new and presumably better concept, the onus is now on Foucault to explain *what* gives discursive practices enough unity that we can talk about *them* even while we cannot talk about "knowledge." Discursive practices are, after all, part of a language of some kind. So it is not immediately clear why this language, if it can express even the most basic bit of mathematics, should be better off than mathematical language. This is important because Foucault himself, in a peculiar way, takes (formalist) mathematics as his model and his dream.

I will show how he does that shortly. First, let us make sure that we understand what is the question that Foucault must answer, for the towering edifice of his philosophy seems to be an impressively elaborate struggle to avoid that question at all costs, notably by burying it under heaps upon heaps of increasingly strange incantations of "the void left by man's disappearance."[7]

The question is simple enough, though: *What* holds discursive practices together, and *what* makes them change from time to time? Obviously, it cannot be language that does that—as shown above by the discussion of Chaitin's theorem, even mathematics, when reduced to language, contains some randomness. Upon such a reduction, there are formulas whose demonstrability (the formal surrogate of "truth") might as well be determined by tossing a coin. This also applies to Foucault's discursive practices. If Foucault were to define his concept properly—which he does not, or does badly—everything that falls under that concept could be simulated by a Turing machine. Deciding where one discursive practice ends and another begins would be just as impossible as guessing the outcome of tossing a coin.

In fact, Foucault cannot avoid this conclusion at all and in the end has to accept that discursive practices change randomly, suddenly, "discontinuously," for no reason other than that they *just* change. But he first tries to explain why this does not seem to happen as randomly as all that in real life.

There is one relatively obvious way to explain this. One could, for instance, say that human beings give discursive practices some inertia but also change them. People keep certain practices in place by means of beliefs, traditions, social institutions, institutions of power, education, politics, indoctrination, control of wealth and media, and so on. People change certain practices through their creativity, errors, speculation, inventiveness, and so forth. This would be, shall we say, a dialectical, romantic-humanist viewpoint. But this is exactly the path that Foucault does *not* want to take. What he wants to do is to import a version of Cavaillès's version of Hilbert, yet taken to an infinite degree, into the human sciences. Let us take a closer look.

First of all, there *is* something that holds discursive practices together. Foucault calls it by various names: the order of things, the *episteme*, power, and so on. It lurks on the outskirts of what we used to call knowledge, which we now call the "discursive practices of an age." It cannot be entirely *in* language, so it is not really a part of these discourses, although it finds its expression through them. It is a subterranean power that makes the formation of concepts possible. But it is beyond our control. Moreover, it is not only formative of humanity; it in some sense determines all humanity because it informs knowledge, that is, discursive practices. We are all, so to speak, thrown into it and depend on it to know anything about anything. To cut the long story short, Foucault's theory seems to be an attempt to formalize that which cannot be formalized, that which Heidegger—Foucault's suppressed source, by the way—called "the mathematical."

We know from chapter 4 that Heidegger understands "the mathematical" as the "space in which we always already move," which makes it possible for us to understand anything *as* an object or *as* truth. Just as it is for Foucault, Heidegger's "the mathematical" is not a construct of individual consciousness and informs the very conditions of our understanding. Heidegger notes that even though "the mathematical" is beyond language, science, and so on, language is all we have to get to it: Language is the House of Being.

Foucault agrees. But in complete opposition to Heidegger, who has serious reservations about formalization, Foucault maintains that "language" is best understood as contemporary linguistics understands it: as a discrete formal structure.[8] From this he draws some conclusions about the human sciences. Human sciences study something called "the human being," but this is not rigorous enough, because—and this is rather obvious—these sciences are studying not *actual* human beings but their *representations* in a language, all the while pretending that they are all about human beings. He proposes that it would be more scientific to leave this romantic illusion behind and, by means of a "return to language," provide the human sciences with mathematical exactitude of which happy positivism always dreamed and which it always envied.

So far, so good. Foucault makes a methodological point, and it is up to the people who work in the human sciences to consider his proposal. I will return to this later, because it has something to do with mathematics. He has also rediscovered Heidegger, which is more interesting at the moment. There is a small problem here for Foucault. Heidegger places emphasis on the notion of continuity, which he claims is beyond formal logic as we know it. What Heidegger calls "the mathematical," we saw in chapter 4, looks like Brouwer's continuum. It is continuous, it has *some* structure, but it eludes at least the accepted ways of formalization.

But this is not the case for Foucault. He remains a staunch formalist and is willing to sacrifice everything to language. Left to provide for their own meaning, formal languages introduce random and inexplicable changes. So be it. Left to their own devices, formal languages seem unable to capture the notion of continuity. So be it. The problem according to Foucault is not in language, but rather in continuity itself. And so, continuity (of time, history, or change in general) is a notion that Foucault will simply discard (thus ending his affair with Heidegger and in fact with the entirety of pre-1966 thought).

At this point we can confirm that the ideological background of Foucault's argumentation is likely to be formalistic at heart. This is nicely shown through a brief comparative analysis in the context of mathematics. For example, intuitionist mathematics was dismissed by "formalists" of all creeds, because it invokes a certain amount of anthropomorphic stuff—even though the unwanted element appears only in disguise, as time-intuition, the continuum, and romantic-intuitionist philosophy of (human) finitude. The idea was not so explicit in Hilbert, and it was Cavaillès who brought it to the fore. But Foucault manages to purify the crudest essence of this objection: He will reject continuity as being an anthropocentric illusion, go after the whole of romanticism, and finally go after Kant.

Apart from this little question of ideology, intuitionism was also challenged on the pragmatic grounds that it offers a strange kind of mathematics in which the greatest achievements of the formalistic approach, as Weyl himself admitted, "dissolve into fog."[9] Foucault repeats this criticism, although in a different form. He challenges the human sciences on the grounds that if they want to be "exact," positive sciences, they have to study formal linguistic structures in a certain manner and give up the claim of being *human* sciences.

The crucial thing, I believe, is that this is not simply a methodological issue, a call for new and better human sciences. Foucault is essentially calling for an end of all thought that gives human beings any special status whatsoever, for instance, by placing them higher than bananas in some unfair anthropocentric hierarchy (an ungrounded "exclusion" and hence an exercise in power).

I conclude that this is the case from Foucault's elaborate descriptions of why human beings are not all that special, other than being a nuisance to his theory, and from his calls for a formalistic "final solution." The human being—*as* something special—is apparently a romantic invention. It appeared somewhere in the late 1770s. This event damaged the notion of certainty and universality of knowledge. Today, which in Foucault's case means the late 1960s, is the time to undo all that.

Let me explain this very carefully, because it might be slightly difficult to absorb. It is not the case, as is usually believed, that romanticism discovered the inability of language to capture humanity. The *opposite* is true. It was due to the workings of language *itself*—or some esoteric power underneath it—that new "ideal elements" arose. These grammatical dummies are what we now call the "human being," the "unconscious," desire, id, inner time, and so forth. They simply came up as an effect of the process of science, much like Cardano stumbled upon imaginary numbers without knowing what he did. Foucault says, in a liberal application of the idea that method produces the object, that "it is not man who constitutes [the human sciences]; it is the general arrangement of the episteme that provides them with a site, summons them, and establishes them—*thus enabling them to constitute man as their object.*"[10]

So, some time in the 1700s "man" was invented, and since then language has been fragmented. If language is to regain its mythical unity, its certainty of communication, and thus bring us back into the state of "felicitous positivism," the narcissistic human being must kindly remove itself from the spotlight of even the human sciences. Since the human being is "barely two hundred years old," its departure will hardly be registered on the scale of the universe. Foucault finds comfort in this thought: "It is comforting, however, and a source of profound relief, to think that man is only a recent invention, a figure not yet two hundred years old, a new wrinkle in our knowledge, and that he will disappear again as soon as that knowledge has discovered a new form."[11] The creature will be erased, as Foucault says, like a face drawn in sand at the edge of sea.

The parallel with mathematical formalism is evident despite Foucault's exorbitant radicalization. The individual human being is sacrificed to abstract formalism so that language may regain its unity. Hilbert, for his part, at least hoped to *prove* that individuals cannot spoil the certainty of knowledge, given a minimal social consensus and the minimum of initial certainty ("real" mathematics). Foucault, on the other hand, hopes that the very concept of humanity will vanish when language chooses to return to its mythical state of primordial unity. He formulates this hope as a rhetorical question: "[W]ill he [i.e., man] not be dispersed when language regains its unity?"[12]

Finally, it seems that almost all the objections that Brouwer raised about mathematical formalism also apply to Foucault, or at least to his followers. To begin with, Brouwer's indictment of the "false belief in the magical character of language" could not be more appropriate in this case. Brouwer also remarked that reducing mathematics to logic would be as strange as "considering the human body to be the application of the science of anatomy." The irony would be lost on a number of Foucault's followers who managed to take his formalism even further: Not only the science of anatomy, but culture in general, or discourse in general, becomes inscribed into the body and thus "constructs" it.

Since we now know where he is coming from, it is not surprising that Foucault's proposed approach is simply to constitute new formal languages of the sort that formalist mathematics already offers in abundance, and will continue to offer in its infinite conceptual becoming. That is exactly what Foucault says in *The Order of Things*, where we can see one of the rare references to mathematics in his work:

> [T]hese questions concern a general formalization of thought and knowledge; and at a time when they were still thought to be dedicated solely to the relation between logic and mathematics, they suddenly open up the possibility, and the task, of purifying the old empirical reason by constituting formal languages, and of applying a second critique of pure reason on the basis of new forms of the mathematical a priori [. . .].[3]

But this seemingly methodological concern has an ideological agenda. For example, even Chomsky's "Cartesian linguistics," which is mathematical through and through, is not good enough for Foucault: Chomsky still maintains a pretense of dealing with the workings of a "sovereign" human mind.

What seems to be at stake is the very idea that people—despite being influenced by discursive practices, which is undeniably the case—can contribute to discourse in some innovative and irreducible manner. That is what Foucault denies when he claims that there is no such thing as "the author." (The credit goes to discourse itself.)

It is relatively clear that there is no deduction involved in such a claim. How can I possibly prove that *no* text could have been authored by a person who—*although* spanned over the endless web of texts that inform her—*nonetheless* contributed to the text in her own irreducible and individual way?

Since Foucault claims that this is not so, and even makes the universal claim that it is never so, one must point out that he cannot support such a universalist position by a rational argument. To prove that all texts write themselves "autonomously," or are "automatically" written in formalistic Ouija board seances, he would have to know all possible texts ahead of time. So the whole thing collapses into a bucket of ideological dust.

Perhaps that sounds too harsh. But Foucault has exerted considerable influence on the "postmodern" sciences of culture, and we have to wonder where exactly he wants to take them. Mathematics shares part of the responsibility, given that Foucault appears to have modeled his notions on a radical version of mathematical formalism. So we have to ask some questions. Here is one: Is postmodernism (whatever that means) in some manner *courting* those extreme branches of science whose objective is to construct a thinking machine?

That seems to be the case. After all, prophets on both sides are summoning a new era. In that era, by its very definition, it would be possible to think without simultaneously thinking that it is a human being that is doing the thinking. That is what Foucault wants, and that is what the apostles of artificial intelligence want.

This is not a coincidence. If one imagines a theory of subject construed by analogy with Foucault's radicalization of the formalist idea, one gets something pretty close to what is popularly referred to as "postmodern subjectivity." This grammatical dummy—complete with its consciousness, body, and the unconscious—is simply an *effect* of mysterious events in a formal languagelike structure.

If such a theory could be supported by a rational argument, the entirety of the artificial intelligence project—in its most extreme form, I might add—would drop out as a surprisingly trivial corollary. More modestly said, the existence of a postmodern subject appears to be logically equivalent to the existence of an intelligent Turing machine. This is exciting stuff, since certain postmodern glitterati claim to be in the possession of the appropriate ontological proof.[14]

Foucault outlines the shape of the human sciences in the era when human beings are entirely replaced by the discursive structures. As a consequence, the whole idea of continuity must be excised. If I assume that science, language, or some other "power," even though it changes all by itself and not through specifically human activity, shows *some* continuity of change, then I could also think that it shows some unity of its own purpose.

Hence, if I so much as think in terms of continuity, say, of history, I am automatically bound up with anthropocentric nonsense. I am projecting my self-image unto something that has absolutely nothing to do with my delusions about my own purposive actions. Since in Foucault's framework human beings are removed from the narrative, arguing the continuity of history would amount to claiming the existence of a divine plan. This is why a truly scientific history can only be an "archaeology," a succession of theatrical acts rather than a movie—discretely ordered acts, staged by forces unknown.

The layers of this "archeology" represent, roughly speaking, a historicized form of the formalist philosophy of mathematics. Where Hilbert relied on the a priori given (but ahistorical) "finitary intuition," Foucault formalizes something that romanticism called the *historical a priori*. In this manner, one gets a discrete sequence of historical configurations, each one being a "meta-theory" for the next. There is no continuity between these archeological layers; they are presented as incommensurable historical facts, which ultimately makes it seem like Kant is the meta-theoretical condition of possibility of the Spice Girls—the epistemic ground on which they boogie. More seriously, the effects of this approach are nicely described by Sartre:

> Each one of these layers defines the conditions of possibility of a certain type of thought which prevailed during a certain period. But Foucault tells

us neither what would be most interesting, namely how each thought is constructed on the basis of these conditions, nor how men move from one thought to another.[15]

Foucault cannot tell us about such things. This is a logical consequence of his formalistic assumptions. I mean this in the precise sense that Foucault's conclusions are an easy corollary of Chaitin's theorem. It is quite simple. *If* our discursive practices are to include a certain amount of formal mathematical language, and *if* actual people are replaced by their formal linguistic surrogates, then there *will* be utterly random ("discontinuous") fluctuations in this formal system. Changes happen from time to time, and that is all that can be said about the matter.

Such an "archeology of knowledge" therefore seems reducible to a single anthropomorphic image. It is the image of a faceless "power" that (once in a while) indifferently tosses a coin to choose a value of the next digit of Ω, thus determining the shape of the new "episteme" or the new "order of discourse."

Foucault is here at odds with Derrida, who always emphasizes the continuity of the historical process, traces ideas as far as our collective memories can reach, and appears to be very far from discarding the notion of continuity altogether. (He calls, rather, for a *rethinking* of it.) This disagreement runs so deep that it in a certain sense makes it impossible to apply the term "poststructuralism" to both Derrida and Foucault.

So there is a split on the inside of poststructuralist theory. The fissure apparently has something to do with the attitude toward the notion of continuity, and in this respect parallels the mathematical "war of factions" over the continuum. The comparison is neither complete nor exact, but every little analogy could be important. I am trying to locate a mathematical "parallel universe" where we might see things more clearly, in a new light and, perhaps, in a less confusing language. Changing the viewpoint sometimes helps.

But what is it that I think we will see in this way? I believe that we might be able to view the theoretical (and to some extent cultural) phenomenon of postmodernism as a curious "product" of the irreconcilable differences between intuitionism and formalism.

We have yet to see how intuitionist arguments enter this picture. At this stage, we can appreciate that in a world where everything is a grammatical dummy there may be problems with the notion of individual freedom. If language (or some other formal structure) constructs everything, our identities and bodies and desires, how can one envision a liberation from this (as Nietzsche wrote) "prison-house of language"?

To resolve this self-inflicted wound of language, some postmodern theorists invoke eclectic variations on romanticism and its successors. In a manner deliberately bordering the sort of discourse that one associates with an unhinged mind, everything from Nietzsche's superhumanism and Byron's poetic flights to psychoanalysis and dadaism is thrown into this verbal salad and tossed vigorously. One example should suffice to get the basic idea across.

The French philosopher Gilles Deleuze—of whom Foucault spoke in embarrassing superlatives such as "one day the age will be Deleuzian"—together with the psychoanalyst Félix Guattari authored a strange book called *Anti-Oedipus* (1972), in which they argue that there *is*, after all, something that precedes formalization. It is not called "the act," the id, Creative Subject, or the inner self. It is called "the savage flow of desire."

The plan for liberating this savage flow from the "Oedipal" or "capitalist" logic that imprisons it is bafflingly simple. It boils down to some sort of ludism in the abstract: absolutely undirected thought, demolishing language with the forceful determination of a salon radical, advocating chaos for chaos's sake, and advertizing delirious, disordered, "horizontal thinking" as the Way of Truth.[16]

The interesting thing, nonetheless (and this is why I introduced the otherwise unremarkable "work" *Anti-Oedipus*), is that the authors strive to explain why the liberated anti-Oedipal thinking will not apply the Principle of the Excluded Middle (the law of "binary" thinking).

One could perhaps say that all those fluctuations and flows of desire look a little like the intuitionist continuum, where the laws of logic are somewhat different. Objects are open, identities flow, movements are always continuous, and the Principle of the Excluded Middle fails for more reason than one. An appeal to intuitionism would have been natural for Deleuze and Guattari, because they start from the romantic-intuitionist premise that there is something that comes before "the sign" (in this case, the "savage flow of desire").

Oedipus remains confined to the binary "mommy-daddy" logic, which suppresses the primordial thinking in continuities. Anti-Oedipus, on the other hand, has broken this "binary code" and lives freely in the continuum (of his desire). So it seems, modulo some terminology, that Anti-Oedipus could have been an intuitionist. Indeed, the authors quite correctly observe that this revolutionary mode of thought is not exactly unheard of. However, according to them, it takes place not in intuitionist mathematics—about which they apparently remain clueless—but in the minds of schizophrenics (in a delirium, of course).

Here is a passage where they attempt to explain what they call the "inclusive use of disjunction" by schizophrenics. This means, I suppose, that Deleuze and Guattari are attempting to illustrate a thinking where the Principle of the Excluded Middle is not applied. Note also the phrase "non-decomposable space," which is the closest these authors come to the description of the continuum. "The schizophrenic is dead *or* alive, not both at once, but each of the two as the terminal point of a distance over which he glides. He is child *or* parent, not both, but one at the end of the other, like two ends of a stick in a non-decomposable space."[17]

Perhaps this will suffice to state my thesis. It seems that mathematics is an important part of the "mommy-daddy" structure that informs the Oedipal drives of popular postmodernism. Sadly, though, the mathematical roots of "high" postmodern theory seem to be well beyond the reach of most of its representatives (with the possible exception of Derrida). This forces the entire edifice to wobble between formalism and intuitionism in a somewhat unstable manner.

It seems to be a symptom. But a symptom of what, exactly? Here is my initial diagnosis. I am afraid that the wobbling is culturally "inscribed" into the mind of our postmodern Oedipus: The infant has a formalist daddy and an intuitionist mommy.

8
SAY HELLO TO THE STRUCTURE BUBBLE

You have a quarrel ... with the algebraists of Paris.
—Edgar Allan Poe

We should take a brief look at the oft-cited source of inspiration of structuralist philosophy, the structural linguistics of Ferdinand de Saussure. This is necessary for two reasons: first, to introduce some terminology, and second, to examine the possibility that the structuralist movement, despite unreasonably frequent references to the work of the Geneva linguist, ended up in a position closer to Cavaillès's formalism than to Saussure's own outlook.

I am, of course, not trying to say that mathematics, formalist or otherwise, bears sole responsibility for various misreadings of Saussure. Things are rarely that simple. The rise of structuralism after World War II was facilitated by a complex set of cultural circumstances, which might even be said to span continents. The victory of formalism and the rebellious siding of young intellectuals with Cavaillès's structuralism against Bergson's and Sartre's intuitionism were among many other relevant factors.

Whatever its formative influences may have been, the fact remains that in the 1950s and 1960s structuralism initiated an uncontrolled reaction of formalization of everything into algebraic-relational structures, a "mathematization" of some areas of the humanities.

In itself, there is nothing wrong with that. Mathematical *modeling* is a crucial part of scientific methodology and probably has something to offer anthropology, literary theory, sociology, and cognitive science. However, as I have already pointed out, uncritically accepting mathematical formalism as a philosophical attitude, especially in its more radical forms, can lead to a fair bit of fashionable nonsense: objects produced by the method itself, languages speaking themselves, and even the notion—my personal favorite, which I discuss in this chapter—of a thinking thermostat.

But let us start with Saussure and, as with Hilbert above, carefully observe the transformation of his ideas in the hands of others.

8.1. Algebra of Language

Structuralism is usually described as the philosophy that renounces the view that single objects can have any meaning on their own. Instead, it holds that the relationship of an object with other elements of some structure is a necessary part of that object's identity. It is through a mutual differentiation among these structural units that we come to assign identity and meaning to them. There is no identity outside of a structural context.[1]

The analogy structuralism is fond of is the example of a crystal lattice. The individual molecules are distinct from one another as well as related (connected) according to the strict rules of formation. These principles of formation would be embodied in the structure (the "blueprint") of the crystal. Our understanding of the elements of the crystal's structure cannot come about by observing single molecules, but rather by noting their relationships in the structure as a whole. We must on the one hand contextualize the elements, and on the other hand distinguish among them within the structural context.

Let us look at another example. In chess, we could easily replace the knight-figure by any other physical thing, say, a cigarette lighter, as long as we understand that the *role* the lighter now plays is governed by the rules that govern the behavior of a knight in chess. Conversely, the physical object we usually identify as the knight-figure can, within a different structure, have a role unrelated to chess. For instance, I could put it under the short leg of my desk to balance the desk properly.

Furthermore, if there were no rules of chess, no physical object could be assigned the identity of a knight in chess, since we would not be able even to formulate the idea of a "knight." In this sense, the structure of chess supersedes both the objects of the game and our ideas about those objects. No physical thing and no concept could be said to articulate anything about chess prior to my consciousness of the rules of chess.

Chess is a relatively good example to keep in mind, but we are now interested in language-in-general as a formal structure. To explain the expression "language-in-general," let us look at the spoken language as a motivating example. Saussure seems to have been concerned with how the meaning of words comes about. How is it that a sound and an idea are united to form a *sign?* Is this mainly a matter of history and heritage? Is it the work of an archetypal father figure of the name-giver—"out of the ground the Lord God formed every beast of the field, and every fowl of the air; and brought them unto Adam to see what he would call them," says the Book of Genesis, leaving us in the dark as to who named all the fish—or can languages be studied in a more rigorous, scientific manner?

It is usually said that prior to Saussure there had been two prevailing approaches to linguistics. One was based on the so-called representational model, according to which the structure of language reflects the structure of thought (i.e., logic). Toward the end of the eighteenth century, perhaps due to the ideas of Hamann and Herder—which I sketched earlier—a break from the representational model occurred.

Henceforth, and this is especially the case with nineteenth-century romanticist thought, language was viewed as a "historical a priori," as something that does not entirely emanate from consciousness but is in a sense prior to it: It is "older than me." Language belongs to a culture, community, history; linguistic actions always include an irreducible element of interpretation; and therefore language cannot be understood solely in terms of the universal laws of logic (thought) as reflected in grammar. For these or similar reasons, a good deal of nineteenth-century linguistics seems to have been concerned with historical, comparative, and etymological studies.

The first of these schools of thought, the "preromantic" one, apparently maintained that there is a *natural* relationship between the idea and phonic expression *of* that idea. The other viewpoint, which is sometimes regarded as having ruled nineteenth-century linguistics, held that the relationship between sound and thought is historically conditioned, and therefore aimed at discovering the etymology of words in a particular language and engaged in comparative studies in order to account for the relationship between the sound and the concept.

Of course, often the same idea corresponds to different sounds, and the same sound could stand for different ideas. Such a variety of mutations, transformations, and redundancies invites us to look for the common traits in these variations, to provide a detailed taxonomy of languages according to their historical roots and other similarity criteria.

But if we are looking for a scientific study of language, something that would move linguistics closer to "science proper," then it must be conceded that collecting data is only part of scientific methodology. The other part usually involves coming up with a mathematical model of the system under consideration. For example, a practical mathematical model of natural language was provided by Chomsky. The method of formal modeling was to some extent neglected in comparative and etymological linguistics, perhaps due to the romanticist aversion to the "determinism" traditionally associated with mathematics.

It appears that Saussure attempted to construct something like a mathematical model of a *general* language: not of spoken language, but of some generic structure that makes language comprehension possible. In this "mathematical" model, there is no predetermined relationship between sound (*signifier*) and thought (*signified*), and hence no sound can convey any meaning on its own. This is a purely formal "language," and we are interested in what can be said about meaning without presupposing *any* historical givenness of the meaning of words. All that matters is what can be said about it in terms of its formal structure.

But such a "language" is perhaps too abstract and far removed from anything that might be of interest to a linguist. For this reason, Saussure distinguishes several levels that constitute an actual language. There is the spoken language, which can itself be viewed in two ways: as *la parole*, the totality of speech-acts of the members of a community, and as *la langage*, the historically evolving collection of actual grammar rules, a treasury of metaphors and "social facts" on which we rely in everyday speech. These aspects of language were emphasized by "romantic" linguistics. In structural linguistics we are looking at something altogether different, namely, the structure that makes language comprehension *possible at all*: an atemporal syntactic blueprint of any language. Saussure calls this blueprint *la langue*. It is a formal structure of linguistic signifiers related in some general manner that we shall consider soon.

However, *langue* is more than a mathematical model of a general language. According to Saussure, it is only through its formal structure that the "material" world of sounds and the world of our thoughts can be brought into a meaningful unity.

Thus, on the structuralist view, the structure of language is a necessary precondition for anything to mean anything. The language-structure serves to articulate the nonphysical semantic component of language (signified) and physical experiences (signifiers). Without it, there would only be the shapeless flow of experiences and an amorphous mass of unarticulated thought. According to Saussure:

> Psychologically our thought apart from its expression in words is only a shapeless and indistinct mass. Philosophers and linguists have always agreed in recognizing that without the help of signs we would be unable to make a clear-cut, consistent distinction between two ideas. Without language, thought is a vague, uncharted nebula.[2]

Speech can become meaningful only through *the very act of articulation* of thought with sound. This articulation occurs *simultaneously*: "Neither are thoughts given material form nor are sounds transformed into mental entities."[3]

The process of articulation takes place within a *general* language-structure. Saussure defines this generic language, *langue*, as a "structure of difference relations with no positive terms."[4] Structural units are assumed to have no assertive ("positive") power of their own, taken in isolation, because this formal structure is deliberately isolated from communal, individual, and historical sources of meaning. These are not discarded, but rather are displaced to the levels of *parole* and *langage*.

But *langue* is what makes articulation possible, and it deserves to be studied on its own. Since the abstract units of *langue* have no "positive" assertive power, the perception of each particular signifier is now the perception of its *difference* from the other signifiers. Grasping the meaning of any one them comes only through gaining an understanding of how it is related to the other signifiers in the entire structure, the role it plays with respect to the *other* units in the structure.

Not all of these ideas are new. The idea that every determination is a negation is not new at all: For Heraclitus, "small" was meaningless without "big," and both Hegel and Fichte emphasized negation, though in a different manner. The structuralist point, I gather, is that we are now not talking about a logical operator—neither a negation *of* something nor a process of negat*ing* in general—but about difference relations that obtain within a formal structure whose units' only relevant feature is that they are distinguishable from one another. This process of determining the relationship of symbols with respect to other symbols in a structure, of determining their roles in the structure, is called structural differentiation.

As for there being no thought without its expression in language, this insight goes at least as far back as the late eighteenth century: It was known to Hamann and Herder. They did not use the word "structure," but romanticists like Schleiermacher and Humboldt did. American philosopher Charles Sanders Peirce—who, by the way, had an interest in mathematics and devised various "logical algebras" as part of his attempt to formulate a general theory of signs—wrote in an 1868 paper that "all thought must necessarily be in signs."[5]

Saussure's contribution seems to have been a careful analysis of what a structure of linguistic signs might look like *in general*, an analysis of the most general formal features of *a* language, independent of its particular realizations at different times and places.

The definition of *langue* as a structure of difference relations with no positive terms encapsulates the notion that language can be studied, to some extent at least, as a formal structure without worrying about the unknown etymological sources of meaning. This does not entail that etymology and comparative studies should simply be discarded. Saussure was himself a Sanskrit expert and spoke several languages.

To employ a mathematical analogy, one can study mathematics in the "formalist" manner, by simply looking at the relationships among symbolic objects in a structure governed by axioms and rules of inference. But this is a convenient abstraction that is somewhat far from practice. If that were how things worked in practice, we would have to prove everything "from scratch" every time. It certainly helps to have some historical awareness of the development of certain notions, their "traditional" meaning, the motivation behind them, and so on. However, for the purposes of a philosophical discussion, one can displace all that to a "meta-mathematical" level.

It is similar with Saussure's linguistics. Let me emphasize this—at the cost of repeating myself—because it is a very important detail. Rather than being discarded, the historical, individual, and communal sources of meaning are displaced by Saussure to the meta-linguistic level, *la parole* and *la langage*, the level of individual acts of speech and communally shared (and historically conditioned) "social facts."

In this manner, *langue* is freed to become, as Saussure famously asserts, a "form and not a substance."[6] It is completely indifferent as to the content of its units. They could denote (as Hilbert nicely said) points, lines, and circles, or tables, chairs, and beer-mugs—as long as they stand in a certain structural relationship.

Now we can formulate the two basic axioms of structural linguistics:

- There is no natural relationship between signifier and signified; the relationship is arbitrary, as long as we are able to differentiate between signifiers.
- It is through *langue* that meaning is articulated; without this structure, our thoughts and experiences would be a shapeless mass.

Let us pause here. It might be interesting to observe that variations on something like the structuralist theme were not absolutely unheard of in mathematics. (I will leave Peirce's logical algebras and theory of signs aside. His work seems to have been unfairly neglected until the 1930s. At that time, perhaps due to the increased cultural significance of mathematical logic, there was a revival of interest in Peirce.)

For example, Frege's Contextuality Principle states that linguistic terms can be considered to have a meaning only within the specific *context* in which they occur and cannot be considered in isolation. Poincaré (who was Saussure's contemporary) and Weyl (who came later) seem to have maintained, in their own terminology, that the relationship between the signifier and the signified is arbitrary, and that mathematicians are primarily interested in the structural relationships among the signifiers:

> Science can only determine its domain of investigation up to isomorphism [of structures]. It remains quite indifferent as to the "essence" of its objects. [. . .] The idea of isomorphism mapping demarcates the self-evident and insurmountable boundary of cognition. (Weyl)

> Mathematicians do not study objects, but relations between objects. Thus, they are free to replace some objects by others so long as the relations re-

main unchanged. Content to them is irrelevant: they are interested in form only. (Poincaré)[7]

It is also interesting to note that the German mathematician Richard Dedekind wrote the following in an influential article from 1888, clearly predating Saussure: "[In] the consideration [of the structure of positive integers] we entirely neglect the special character of the elements; simply retaining their *distinguishability* and taking into account only the *relations to one another* in which they are placed [. . .]."[8]

Finally, the idea of studying purely structural relationships among the "meaningless" ideal elements of an abstract structure—elements that get their "meaning" only when they are articulated within a formal demonstration (i.e., proof)—was officially inaugurated by Hilbert, first in his *Foundations of Geometry* (1899), and henceforth in his views of mathematics in general.

I am, of course, not attempting to belittle Saussure's work by making it look like he was lifting ideas from mathematics. That would be nonsense, and I have no intention of trying anything of the sort. On the contrary, I am searching for common patterns that could serve as the ground for developing analogies and comparisons as to the "fate" of both structural linguistics and mathematics.[9]

We saw in the preceding chapter that Hilbert's formalism ended up having difficulties when it tried to suppress the specifically human sources of mathematical meaning. They were initially displaced into "meta-mathematics" but were then dismissed altogether when finitary intuition acquired the shape of an idealized machine. A similar transformation occurred in linguistics, and we will have to see how.

Before I try to develop this mathematical analogy in more detail, let me say that the germ of the structuralist idea can be traced to the early nineteenth century, to Schleiermacher and Humboldt. Schleiermacher, for example, explicitly used the term "structure" in what could be regarded as quite contemporary manner. By "structure," he meant a system of relations among elements whereby each element derives its meaning through the unequivocal differentiation from all other elements. And he observed that such a definition of "structure" could be applied not only to language, but also to cultural, social, economic and juridical orders.[10]

These historical observations are important. They demonstrate that "structuralist" notions were lurking on the cultural horizon and were articulated with more or less clarity by people of *very* different philosophical outlooks. Schleiermacher, Poincaré, and Weyl belong in one way or another to the romanticist tradition. Dedekind and Frege were rationalists. Hilbert had little sympathy for Poincaré's and Weyl's "romanticist" views. But how are such ideological differences reflected in the variations on the structuralist idea?

Practically everyone (with the possible exception of Brouwer, who made some eccentric claims about mathematics as a "languageless activity") agreed that formal structuring of signifiers is a *necessary* component in the creation of meaning. However, there is considerable friction regarding the converse. One can speculate with a good deal of certainty that Weyl and Poincaré would not have agreed that purely formal differentiation among the signifiers *suffices* for the creation of meaning. It is perhaps unnecessary to point out that Schleiermacher and Humboldt, whose views rested on the primacy of the individual act of interpretation, were of the same opinion. Hilbert himself, although his school later turned everything into a formal structure, started out by saying that finitary intuition—our capability to survey the units of a symbolic structure in (as Hilbert says) "their difference, their succession"— is a *pre*condition for the activation of all formal inferences.

We are coming close to the point where "structuralism" splits into two incompatible factions. Let us look more closely at what Saussure's opinion might have been.

According to Saussure, the meaning of the linguistic sign (the couple signifier/signified) occurs through the act of simultaneous articulation of thought and sound. In this process, various structural units are differentiated from one another and produce a "value" by virtue of being different from the other units. The actual *event* of articulation, however, takes place on the level of speech-acts, that is, *parole*. As such, it steps outside the atemporality of *langue*. It acquires a uniqueness in space and time. It becomes subject to the historicity of *langage* and *parole*, the "social facts" and individual acts of speech that are absent from the formal structure of language.

For instance, the same sentence—"this cubic equation has a solution," or, to use a more dramatic example, "Madonna is not a virgin at all"—would be interpreted quite differently in the sixteenth and the twentieth centuries. That, you might say, is quite obvious. Indeed. But it is nevertheless true that Saussure's numerous "followers" held that *langue itself* determines all meaning (or at least some formal surrogate of meaning). That seems closer to Cavaillès's ideas about science speaking to itself all by itself. A little later, we will look at the apparent misunderstanding that ensued and materialized itself as the intellectual phenomenon called structuralism. For now, let me emphasize the point where the subsequent ideological coup would be carried out.

Here is the question that we must answer: Can it be said that the formal structural differentiation within *langue* is not only *necessary* but is also *sufficient* to fix the identity of the linguistic units?

There are people who seem to think that the answer is in the affirmative. But the issue is far from settled. Frege and Dedekind, for example, encountered difficulties in trying to prove such a claim in the case of the "*langue*" of arithmetic. Frege ended up with a paradox on his hands. (It was discovered by Russell in 1901.) Dedekind chose the privacy of a letter to relate this explication as to how it is that we comprehend the structure of positive integers: "[W]e are of divine kind." So it is not at all obvious that formal structures *do* fix the meaning of their units, even if the meaning cannot be fixed *without* these structures.

Indeed, for Saussure, formal structural differentiation is necessary but *not* sufficient. Here is what he says about the identity of structural units: "This identity always includes an unidentifiable subjective element." Furthermore, "there must be the first act of interpretation, which is active [. . .]."[11] This sounds much closer to the romanticist tenet that "in the beginning there was the act" than to the formalism that was later attributed to Saussure.

It appears, then, that Saussure did not mean to say that *langue* was a universally given virtuality that informs every thought of every human being in a mystical yet deterministic fashion. For Saussure, forms and grammar exist socially, but changes arise from the individual.[12]

The question must now be asked: How did it happen that Saussure's name became associated with the view that "language determines thought"—and even "determines" it to such an extent that the whole concept of authorship, relative autonomy of individual meaning attributions, and other anthropocentric junk must be sacrificed at the altar of formalism and dissolved in the structures of *langue*, Turing's machinery, discourse, and so forth? How did the structuralist movement make the giant leap from Saussure's "there is no thought without language" to things like "language determines thought," "language speaks," and "there is no author"?

The story is a complicated one, as are most historical narratives. I can only hint at a few possibilities, and paint the cultural landscape in broad brushstrokes that appear to make sense in hindsight.

First of all, recent critical studies have shown that the editors of Saussure's seminal book, *Course in General Linguistics*—which was assembled from his students' lecture notes and published posthumously in 1916—may have taken certain liberties. This might help explain part of the confusion. But it still seems somewhat farfetched to claim that a single work published in 1916 suddenly inspired an entire philosophical movement 40 years later. That does not seem to be how things work, and there is no reason to believe that books written by Sanskrit experts from Geneva should somehow be privileged in this respect.

I have already noted that the extreme formalism of Cavaillès's philosophy of science may have had a role in the rise of structuralism. That was certainly a relevant factor. But there were other equally interesting and independent factors. For example, the structural anthropology of Claude Lévi-Strauss, which we will consider briefly in a mathematical context, also played an important role.

Lévi-Strauss was concerned, among other things, with the structure of myths. His theory is rather general and therefore applies also to the myth of immutable mathematical truth: the belief that theorems, no matter *when* they are actually proved, must *always* have been true. This myth undoubtedly operates in the mathematical community, so let me try to sketch out the basic ideas of structural anthropology as applied to the ancient culture of mathematics.

To begin with, let us say that Hilbert's "ideal mathematics" is a system that in a sense parallels Saussure's *langue*. It governs the structure of thought and ideal judgments. The ideal elements produced by this structure have no "positive" value on their own. Also, the relationship between the signifier and the signified is arbitrary in formalist mathematics. The signifiers, ideal elements like the square root of -1, only become "something" through the act of demonstration ("articulation"). So there are some similarities between Hilbert and Saussure. The analogy may of course be open to some objections, but I promise I will not try to run too far with it. I am only using it as a convenient narrative trick to illustrate what Lévi-Strauss made of Saussure's linguistics, while keeping our earlier discussion of formalism present in the background.

Now suppose I managed to prove a theorem, for example, a theorem of geometry, considered as a formal structure the way Hilbert formalized it in *Foundations of Geometry*. This formal structure is part of the *langue* of ideal mathematics. But my demonstration—just like Saussure's "articulation" of sound and thought in an act of speech—occurs in actual space and time. I do not live in the ahistorical universe of ideal mathematics. So the proof takes place in real life, on the level of real mathematics, which in some sense corresponds to Saussure's *parole*. This might create some problems for a formalist.

For instance, I could have interpreted the signifiers in the formal system of geometry as standing for tables and beer-mugs, which *then* motivated me to formally prove the theorem. In this manner, I could have made some "unidentifiable subjective contribution" (to use Saussure's expression). This contribution marks my act of proving the theorem as different from some other demonstration of the same theorem. The same theorem could have been proved by someone on a Greek island many centuries ago. The two proofs could even be formally identical as sequences of symbols, but they still seem to be different in *some* sense, namely, as actions of different human beings with possibly different motivations, occurring at different times and places. Therefore, the "truth" of this theorem, *if* it is identified with the *act* of demonstration, might still be marked by temporality and individuality of the proof's author.

Naturally, it would be very strange to claim that Pythagoras's theorem is true because *I* demonstrated it this morning. It is true whether I proved it or not. Hence, theorems are in one sense acts of demonstration and as such reside *in* time, but they are in another sense timelessly true (i.e., are believed to be so).

Let me recapitulate. If the truth of a theorem is *in* the demonstration itself, then the truth of that theorem, strictly speaking, eludes the ahistorical *langue* of ideal mathematics. Yet most mathematicians believe that theorems are timelessly true, *as if* theorems were part of some ideal language that is oblivious to history. Theorems follow one another in time, but this fact is somehow dissolved in the myth of their timeless truth.

So it seems that mathematics unfolds not in "usual" time, but in what Lévi-Strauss calls "mythical time." Myths, as actual narratives, occur in real life, on the level of *parole*: Someone tells a story. But they are nevertheless "timeless." They are stories that are believed to be eternally true.

From a similar analysis applied to Saussure's linguistics—I believe that I reproduced the argument faithfully—Lévi-Strauss would make the following conclusion. In addition to the two levels of real and ideal mathematics, analogous to the "real" *parole-langage* and "ideal" *langue*, there is yet a third level: that of "discourse." (This is how the term "discourse," in its contemporary sense, popped into being.) The structure of discourse is the structure of myths, in our case the myth of mathematical truth.

So far, so good. Lévi-Strauss introduces another structure, a superstructure of some kind. He wants to study this structure in a scientific, logical, even mathematical manner. How would this be done? Precisely as in formalist mathematics. Recall that Hilbert sought to remove the ambiguities associated with the fact that proofs are carried out by actual people. That would be one way of understanding Hilbert's refusal to deal with motivations, interpretations, "unidentifiable subjective elements," and other "anthropocentric" nonsense. *Actual* demonstration of a theorem, the meaning attributed to it by the people who demonstrated it, the history and the motivation behind the proof, all of that is suppressed. What matters to Hilbert is whether the theorem *can* be demonstrated at all ("in principle").

Similarly, it does not matter to structural anthropology which particular mythical narrative we are considering—geometry, kinship, love—or when, how, and by whom it is actually narrated. All these narratives are produced by the same "method," which resides in the "collective unconscious."

So there is this timeless matrix that generates all myths. Now *it* must be studied, in the manner that parallels the formal methodology of mathematical logic. That is what Lévi-Strauss says: "[T]he logic of mythical thought is as rigorous as that of modern science, [. . .] the difference lies, not in the quality of the intellectual process, but in the nature of the things to which it is applied."[13] Thus, the structure of myths, "discourse," will have to be *idealized*, just as one does in modern science. It has to be given a formal model, if structural anthropology is to have the desired rigor of logic.

This will create a familiar difficulty. To see what shape it takes in this context, let us grant that we have this ultimate structure, "discourse," the structure of myths. This is an abstract, formal structure of some kind. What, if anything, provides meaning to the units of this formal structure? The problem does not occur in Saussure's linguistics: His scheme permits individual subjective contributions, social facts, and so on, all of which reside on a "meta-level" that he called *parole* or *langage*.

But there is no "meta-level" for Lévi-Strauss, not any more. The situation is roughly analogous to Hilbert's formalism. Finitary intuition, the "real" background of formal systems, became idealized as a Turing machine and thus became part of some timeless, abstract structure. At that point the question arises as to how these idealized formal systems—once their "real" carpet has been pulled from under their "ideal" feet—can have any meaning at all. Recall the question that Gödel addressed to formalism: How could anyone think of expressing meta-mathematics *in* formal systems, if the latter are considered to consist of meaningless symbols that acquire some substitute of meaning *through* meta-mathematics?[14]

A slight variation of the same question applies to Lévi-Strauss's structure of discourse. It may be that there is such a structure, and that it resides in the "collective unconscious." It may even be possible to find a nice formal *model* for this structure. But suppose that this model is proclaimed to actually *be* the structure. When this is done, the structure will become part of the *langue* of ideal mathematics. This may well short-circuit the whole project. Such formal structures—for Saussure, and initially for Hilbert—were supposed to get their meaning from a "real" level of language, which includes all the acts of formal demonstration, linguistic articulation, and narration of stories: acts performed in real time by real people. Yet Lévi-Strauss claims that all of these "real" activities are *in fact* part of the great structure of "mythic discourse" and therefore reside *in* an ideal, purely formal structure. So how will this structure mean anything? (And to *whom?*)

A little bit of magic always helps. This ultimate structure provides, somehow, its own meaning. It does not mean something because a human being realizes her potential in some unifying interpretive act, the act of narrating a myth or proving a theorem. Just the opposite is the case. It is the structure *itself* that speaks (speaks to itself, I guess) through human beings. We are already familiar with this notion from Cavaillès's philosophy of mathematics. To be sure that this is the opinion to which Lévi-Strauss subscribes: "I therefore claim to show, not how men think in myths, but how myths operate in men's minds without their being aware of the fact."[15]

Let us see where we stand now. We have reached the ultimate level of idealization, the structure of all structures, as it were, and there is hardly a way upward. The thing has to provide for its own meaning, because there is no meta-level that it could invoke. It cannot appeal to people, since they are mere conduits for the structure's divine plan. We have structuralized and idealized all there is. As for *how* this creature is supposed to provide for its own meaning, the explanation follows this simple pattern: Structural differentiation among the ideal units produces the meaning of the units all by itself. Structure is not only necessary for the creation of meaning, it is also sufficient.

Semantics is in the syntax. (The corollaries are numerous, and I will outline some of them in the next section.)

With this, structuralism comes very close indeed to the most staunchly formalistic branches of science. In fact, the notion that "if you take care of the syntax, semantics will take care of itself" is occasionally found in essays on cognitive science and artificial intelligence.

This idea has little to do with Saussure, but there are interesting structural similarities between Cavaillès's understanding of the term "structure" and some of Lévi-Strauss's own assertions. For the former, the "science-structure" develops of its own internal necessity that is independent of the intentions of individuals. It rolls forward like a Hegelian runaway train, producing ever higher levels of reflection in an infinite effort to convince itself of its own consistency. For Lévi-Strauss, something similar is true of myths: "[S]ince the purpose of myth is to provide a logical model capable of overcoming a contradiction [. . .] a theoretically infinite number of slates will be generated, each one slightly different from the others."[16]

For both of these thinkers, the intentions and desires of individual human beings are irrelevant. Both of them agree that there is something larger that speaks through me when I prove a theorem or tell a story. Both of them had an aversion to Bergson's "intuitionism," to phenomenology, and to existentialism. And both were impressed by formalism of one kind or another.

All of this seems to have played a role in the transformation of Saussure's linguistics into something that Saussure most likely did not mean at all. There are probably numerous other factors, such as the role of migrating Russian and Central European scholars, but I cannot hope to give a complete account of the relevant cultural circumstances. Let me mention, however, the influence of Nicholas Bourbaki.

Bourbaki is a fictional character who nevertheless exerted considerable influence on the twentieth-century mind, thereby confirming Herder's hypothesis that mind is shaped by fictions. Nicholas Bourbaki is a pseudonym of a group of French mathematicians who were apparently so taken with the structuralist idea that they felt compelled to form the Bourbaki entity as a collective consisting of anonymous individuals.

Immediately upon his birth in the 1930s, this "man" undertook the monumental task of rewriting all of mathematics in terms of abstract structures. Bourbaki's views to some extent paralleled the pedagogical theories of the Swiss physician, psychologist, and educationalist Jean Piaget—there was eventually a collaborative effort between them—who considered the development of cognition to be inseparable from the acquisition of operational knowledge (acquired "intuition") of certain mathematical structures.[17] The curriculum reform that resulted from the efforts of such lobby groups is represented in the familiar phenomenon of "new math." (It is called "modern mathematics" on the continent.)

I have intentionally heaped up all these developments from the 1930s and the 1940s, in order to illustrate the possibility that mathematics may have had a role in the rise to fame of the word "structure." Such cultural phenomena were described by the American mathematician Gian-Carlo Rota as "the pernicious influence of mathematics on philosophy." A similar opinion was expressed by the French mathematician René Thom, whose view of Bourbaki's influence on the mathematical curriculum is nicely summed up in the title of one his polemical articles: "Modern Mathematics: Educational and Philosophical Error?" (1971).[18]

This is not to say that mathematics is the sole culprit. But from this perspective, the ensuing structure-bubble would appear to have more to do with the influx of a new and hence exciting "mathematical" formalism into the (allegedly) methodologically challenged social sciences, than with Saussure's linguistics. It may be that it was by the powerful engines of this fashionable patois that structuralism was propelled toward the "postmodern" generation.

In this sense, one may well say that our "postmodern" Oedipus has a formalist "daddy." In the next section, I will try to indicate the point at which poststructuralist critiques of structuralism are addressed, but it is only toward the end that we will be able to examine the argumentation that poststructuralism seems to have borrowed—or at any rate could be seen as having borrowed—from its intuitionist "mommy."

For now, we should note that the structuralist movement enters a phase of self-critique, even if only in the form of some apparently insignificant Freudian slips, already by the mid-1960s. For instance, the French writer and culture-theorist Roland Barthes wrote in his 1963 essay "Structuralist Activity":

> *Structure* is already an old word [. . .], today quite overworked: all the social sciences resort to it abundantly, and its use can distinguish no one [. . .]; *functions, forms, signs, significations* are scarcely more pertinent: they are, today, words in common use from which one asks (and obtains) whatever one wants, notably the camouflage of the old determinist schema of cause and product [. . .].[19]

Let me introduce a terminological convention to avoid unnecessary confusions between Saussure and his alleged followers. Let us reserve the word "structuralism" for something that may be closer to Saussure's line of thought, namely, the view that structural differentiation is a necessary but *not* a sufficient condition for the occurrence of meaning. As for the other kind of "structuralism," where meaning is supposed to spring out of the structural differentiation alone, I propose that it be denoted by the term "functionalism."

Let me try to explain why this seems to fit it better, and consider some consequences of this outlook.

8.2. Functionalism *Chic*

Let us assume, for the sake of illustration, that we are given some purely formal structure of the kind discussed above. Since a formal structure can be applied to almost anything one can think of, it is best left unspecified. I will simply speak of some generic structure and its elements.

According to the functionalist view, semantics is mechanically inferred from syntax. The elements of the structure get their meaning from their mutual interdependencies according to the rules of the structure, period. As a consequence, the elements encountered in a structure become functions. Their meaning is identifiable with the "roles" they play in the entirety of structure. Their material expressions (or any other characteristics that are independent of the role in the structural formation), as well as the "observer's" intentions, are simply irrelevant. Anything can be a symbol. The relationship between the signifier and the signified is arbitrary.

This may already be clear from the chess example above. The notion of "knight" derives its meaning from the rules of chess, while the actual knight-figure can be replaced by anything. "Knight" is a function of the chess structure. It gives its name to a *role*. What we have, then, is a *pure function*. The set of these functions constitutes some semblance of a semantics, which we, should we be so lucky, might *discover* (as opposed to partaking in its creation, which is forbidden). Functionalism, in this sense, appears to be a species of Platonism.

But it would be somewhat simplistic to claim that it is entirely reducible to this ancient doctrine. It is the greatest triumph of abstract formalism that it can put under the same functionalist umbrella such diverse thinkers as Nietzsche—who believed that all thinking proceeds under the spell of certain grammatical functions—and computer scientists who claim that if you take care of the syntax, semantics will take care of itself.

That is quite impressive. Let us consider some corollaries of functionalism, in order to get some sense of what else might fit under this general rubric.

Corollary 1. Functionalist Literary Theory

There are many examples of this influential trend. It seems, for instance, that Lévi-Strauss's functionalist anthropology might have been inspired by it, through his acquaintance with a Russian literary theorist with whom he worked during the 1940s in New York. Vladimir Propp, another Russian literary theorist, gave a structural analysis of Russian folktales. He concluded that all folktales have the same basic structure and that events they talk of are only "functions" determined by their role with respect to other "functions" in the structure of the tale. This, in turn, influenced the work of the

Lithuanian-born French semiotician Julien Algridas Greimas, who actually went so far as to talk about *the total meaning* of a literary text.[20]

Corollary 2. Functionalist Theory of Language and Culture

With a slight stretch of imagination, community can be considered a culture-structure that imposes certain linguistic rules (conventions) on its elements (people). Functionalist explanation of the semantics of this structure now entails linguistic determinism. This idea has a curious history and excites the minds of many to this day.

In the 1940s, the idea of linguistic determinism was being popularized by the American linguist Benjamin Lee Whorf. The so-called Sapir-Whorf hypothesis, in its strongest version, claims that language determines thought. It also implies the thesis of cultural relativism, which states that cross-cultural communication is uncertain ("the impossibility of translation").

In a profusely quoted article from 1940, Whorf says:

> We cut nature up, organize it into concepts, and ascribe significances as we do, largely because we are parties to an agreement to organize it in this way—an agreement that holds throughout our speech community and is codified in the patterns of our language. The agreement is, of course, an implicit and unstated one, *but its terms are absolutely obligatory.*[21]

There are two ways of understanding this claim. One would be that language informs our thinking by providing the treasury of metaphors, "social facts" handed down through tradition, which we invoke when expressing ourselves. This viewpoint has been in circulation at the very least since the time of Herder, who said that mind is formed by fictions. It is dependent on the "historical a priori" of language, which is "older than me." From Herder onward, this view extends throughout the nineteenth century (and can also be sensed in Saussure).

This informative role of language implies cultural relativism, in the sense that absolutely certain communication between different linguistic communities cannot be guaranteed. I noted in chapter 2 that this was Herder's idea, which he furthermore supplemented by saying that even the agreement of a single community remains unstable and has to be reconfirmed in practice.

So *that* version of the Sapir-Whorf hypothesis is, to say the least, not headline news. It is a complete commonplace of all romanticist philosophy. The curious bit is the strong version, which claims, as Whorf apparently does in the passage I quoted, that the "historical a priori" of language is *absolutely obligatory*. In this case language not just informs thought—as it does in romanticism and in Saussure—but actually determines it.

Since the mid-nineteenth century, variations on this theme have been employed as a sort of ideological weapon. For instance, Karl Marx wrote in the preface to his *Critique of Political Economy* (1859) that "it is not the consciousness of human beings which determines their being, but it is, on the contrary, their social being that determines their consciousness." (Also, Chairman Mao's *Little Red Book*, a source typically passed over in silence in this debate, should not be underestimated for its influence on a number of key "postmodern" thinkers: "It is man's social being that determines his thinking.")[22] A predecessor of the thesis of linguistic determinism can also be found in other parts of the ideological spectrum, for example, in Nietzsche's assertions to the effect that thought always proceeds under the domination of certain grammatical functions.[23]

The hypothesis of cultural-linguistic determinism seems quite extreme and difficult to test. An entertaining critique of this notion by a contemporary linguist can be found in Steven Pinker's *The Language Instinct*.[24] Nevertheless, the issue seems to be a matter of some controversy, and a passionate debate among various interested parties is still in full swing. I mention it only as part of the complex web of circumstances that led toward the notion that some "formal" structure—society, political economy, culture, myth, language, grammar, and so on—alienates what romanticism regarded as an unalienable human right: the right to attribute meaning to things in an individual (but not completely culture-independent) manner.

Along with these theories of linguistic and cultural determinism, a theory of the media based on functionalist semantics has been expounded by the Canadian scholar Marshall McLuhan. In his case, the familiar idea that "language speaks" or "structure speaks" surfaces in the form of technological determinism, especially with regard to the media.

McLuhan is the inventor of the well-known mantra "the medium is the message." We are now in a position to understand what he means. Individual "messages" are arbitrary physical representatives of certain units within the structure of media. The relationship between the signifier and the signified is arbitrary, and hence the "programming content" is arbitrary. Each unit is meaningful only as a function of the entire system, and thus spreads the messianic message of the medium *itself* (and not of the people who consider, say, MTV, to be a continuous marketing event).

Let us see what else is covered by the surprisingly comprehensive policy of functionalism.

Corollary 3. Functionalist Theory of Mind

Some such theory, with more or less sophisticated variations, is at the root of the belief that the human mind is functionally equivalent to a machine. The theory holds that the human mind is a syntactic structure whose functionalist semantics consists in its system-states. That is to say, the physical nature of the network of units that comprise the mind-hardware is irrelevant (the relationship between the signifier and the signified is arbitrary). The identity assigned to the structural units is a function of their relationship with the rest of the system; that is, it becomes a "function" of the system as a whole, and it becomes that all by itself, through structural differentiation (plus the stimulus of the environment).

An interesting example of functionalist semantics (in the sense in which I am using the term) is provided by the American computer scientist John McCarthy—also known for coining the term "artificial intelligence"—who famously maintains that a thermostat could be said to have "mental states," or beliefs (i.e., it has its own semantics). A thermostat, according to McCarthy, has three mental states: "it's too cold in here," "it's too hot in here," and "it's just about right in here."[25]

Corollary 4. Functionalist Hinduism

A strange idea, I admit, but I cannot resist coming up with my own functionalist theory. Say that the nonphysical world of our thoughts is analogous to the nonmaterial world of souls, and that the physical world is analogous to the collection of physical bodies of all living creatures. Now I can formulate my functionalist take on Hinduism.

The world-of-souls is what would be called Atman. Atman freely transmigrates from body to body. We have all heard people speak about the "transmigration of souls." That is a vague, unscientific expression. From the functionalist standpoint, we should rather say that the relationship of the signifier and the signified is arbitrary. That is much better. Furthermore, Hinduism sometimes describes our special quality of having a soul as "Atman-embodied-in-a-physical-body," just as thought and sound remain shapeless until language articulates them. Thus, the concept of articulation replaces the coarse-sounding "incarnation."

Now we have fixed some impressive terminology, but that cannot be all. Things cannot go on as arbitrarily as all that. If I am, for instance, a well-known American physicist, it would be strange to wake up one morning only to find that "my soul" is now embodied in the physical body of, say, a French feminist poststructuralist. It could be interesting, no doubt. But somehow it does not seem right. To complete my functionalist account of Hinduism, I must come up with some structure to govern all this. Luckily, in Hinduism, there is the concept of dharma. So dharma is the structure, and my "role" is to abide by its rules, through which I establish my relationship with other bodily signifiers. Otherwise, I collect bad karma and make things worse for the old soul. Q.E.D.[26]

For all its entertainment value—one can hardly ask for more than to see a thinking thermostat—it is perhaps time to note that there seems to be a bit of a problem with functionalism. The problem is that the "meaning" of each element of some formal structure is identified with its role with respect to all other structural units. The meaning of each element of the structure is defined in terms of the totality to which it belongs. Thus, the "identity" that structural differentiation is supposed to assign to the units of the structure involves what Poincaré called nonpredicative definitions (now known as "impredicative"). It is therefore open to the same objections that Poincaré addressed to logicians and formalists, back in the days when mathematicians were in the business of "deconstructing identity."

I indicated in the discussion of Poincaré—and I will revisit this objection as elaborated by Derrida—that unlimited use of impredicative definitions may lead to difficulties with the "mutability" of the semantics of structures with unbounded numbers of elements. This is certainly the case with any language-structure that is supposed to be capable of expressing elementary assertions about numbers. It seems to be a general feature of such definitions that upon extending the structure, the "function" assigned to some unit could in principle change. There is no way to predict exactly how it would change.

On the other hand, the structures in which the functionalists are interested are typically capable of creating new structural units: new words, sentences, myths, people, endless arrays of sitcoms, and so on. Therefore, the meaning that the functionalist assigns to these units cannot be a priori guaranteed to have any permanence or "continuity." The situation is serious. If it is the structure itself that "speaks" through us, if it is the structure itself that provides for its own meaning, then the community of human beings cannot help it remain stable. We are utterly powerless with respect to this ineffable virtuality, and we can no more prevent it from changing than we can institute changes in it.

At this point a new assumption therefore comes into play, the assumption of stability, yet another mysterious principle that guarantees that the structure will not deteriorate like a spider dissolving into its own web. The principle that functionalism must invoke has been expressed very nicely by Piaget: "These properties of conservation along with the stability of boundaries despite the construction of indefinitely many new elements presuppose that structures are self-regulating."[27]

It is precisely at this issue that poststructuralism aims its fundamental objection: Why, exactly, should this principle of self-regulation be regarded as more than a convenient or even arbitrary *assumption*? How can the structure be guaranteed to be self-regulating? It is through this question that certain "postmodern" arguments can be related to Poincaré's critique of impredicative definitions.

Before we consider that analogy in more detail, let us glance at Wittgenstein's later work.

9
DON'T THINK, LOOK

Yo soy yo y mi circunstancia.
(I am I and my circumstances.)
—José Ortega y Gasset

Wittgenstein's work is traditionally divided into two or three phases. His early period is related to logical positivism and probably reflects his interest in the ideas of Russell and Frege; it is separated from the subsequent periods by a long hiatus. His return to active philosophical work, which marks the entrance into the middle phase, appears to have followed Brouwer's lectures in Vienna in 1928.

Indeed, Wittgenstein's later work explicitly takes up some of the issues mentioned by Brouwer in his lectures, for example, the role of "training the mind" and culturalization of individuals (Wittgenstein had been a schoolteacher and already developed an interest in these problems), and transmission of the will by means of what Brouwer called a "single human cry" (Wittgenstein discusses utterances such as "Sit!", "Water!", etc.).

This is not to say that the two were in complete agreement, and I will try to track down their differences soon enough. But it seems safe to say that Brouwer's lectures on the philosophy of mathematics, an area of study that had attracted Wittgenstein to philosophy in the first place, may have played a role in inspiring his return to philosophical activity.

Let us look at two important questions that connect Brouwer and Wittgenstein.

The problem of following a rule asks whether any rule can determine a particular course of action. This has something to do with Brouwer's claim that no language can ensure absolutely accurate transmission of the will. "There is no certainty in will-transmission," wrote Brouwer, "especially not in will-transmission by language." Brouwer was particularly interested in the case of the intended meaning of a mathematical rule. In his Vienna lecture, he said: "There is, therefore, also for pure mathematics no certain language."[1] Wittgenstein reworks these remarks and supports them by numerous examples, mathematical and otherwise.

The private language problem asks whether an individual could construct a meaningful language entirely on her own, without invoking external criteria. Much has been written about the private language as a language of "sensations" (e.g., pain). But for Wittgenstein, sensations present only one possible aspect of some would-be private language. The problem has a history going back at least to Nietzsche.[2]

But in the present context, I am interested mainly in its relation to Brouwer's extraordinary claim that mathematics is an essentially languageless activity of the self, an activity that cannot in any manner be reduced to mathematical language. For Brouwer, there is no such thing as a private *language*—"language is a function of social man"—but there *is* something private, some accompanying activity of the mind, which eludes language. (He calls it by various names: the will, mathematical attention, the self, the creative subject; all of them are "far from reasoning and words.")

Wittgenstein definitely accepts the *first* part of this claim. However, things are not so simple with regard to that private "remainder" of language. Wittgenstein wants to reject this notion along with Brouwer's extreme solipsism. The basic idea is that there can be no meaning without language, and language is social, so there can be no "private meaning." Thus, Brouwer's solipsism ends up reduced to nonsense. I cannot doubt the existence of the "outside world" or of "others" if nothing can mean anything without them.

So far, so good. Wittgenstein debunks Brouwer's strange notion that mathematics is an "essentially languageless activity," along with the claim that other people are sequences of Brouwer's thoughts. I will argue below that Wittgenstein could have simply quoted Fichte to establish *that* much. The problem is that if Wittgenstein's argument is taken literally, as saying that there is indeed no private meaning at all, then it seems that he is arguing something unpleasantly close to cultural determinism. If meaning is strictly public, then the role of the individual is reduced to parroting back various cultural conventions. That would bring Wittgenstein—and other theories of language that rest on this view, including parts of postmodern theory and parts of Anglo-American language philosophy—to the ideological vicinity of various theories about the "inscription" of culture into all individuals ("branding" of the "herd," to use a more Nietzschean mode of expression).

That is a self-defeating conclusion that Wittgenstein himself does not seem to trust. His resolution of the difficulty is somewhat sophistic. He appeals to pragmatism to reject the practical *utility* of the romantic-idealist notion that a part of the self cannot be captured in language—*because* it cannot be captured in language. ("It divides out, whatever it is.") Nonetheless, he repeatedly finds himself nodding at romanticism, to be sure in an oblique fashion, throughout his *Philosophical Investigations*. Let us look at a couple of examples.

Recall that Schelling criticized Descartes's "I think, therefore I am" (*cogito ergo sum*) as failing to prove what Descartes thought it would. According to Schelling, Descartes's assertion does not prove that I unconditionally *am*. At best, it establishes that I am in a particular *mode* of being, that I am-*as*-a-thinking-being (which therefore does not capture my full being with any more certainty than the less impressive "I produce fluids, therefore I am in the bodily mode of being"). Schelling's exact wording: "The *sum* that is enclosed in the *cogito* does not, therefore, have the meaning of an unconditioned 'I am,' but rather only the meaning of an 'I am *in a particular way*,' namely, as thinking, as being in the mode that is called thinking."[3]

Wittgenstein, in paragraph 417 of *Philosophical Investigations*, basically repeats this remark: "[T]he sentence 'I perceive I am conscious' does not say that I am conscious, but that my attention is disposed in such-and-such a way."

Here are a few more nods of this kind. In the late eighteenth century, Hamann expounded the view that the body is the image of the soul.[4] In the early nineteenth century, Schleiermacher wrote that every act of understanding has the character of a work of art.[5] In *Philosophical Investigations* (II-iv) Wittgenstein famously claims that "the human body is the best picture of the human soul," and in paragraph 491: "Understanding a sentence is much more akin to understanding a piece of music than one may think."

Finally, although the detail is occasionally passed over in silence, Wittgenstein notes in a truly circumspect manner that I may in fact be at liberty to use my imagination to produce some semblance of meaning for myself. In *Philosophical Investigations* (II-xi, p. 210), he describes an encounter with a strange new symbol he had not seen before and notes: "And I can see it in a variety of aspects according to the fiction I surround it with. And here there is a close kinship with 'experiencing the meaning of a word.'"

Let us ignore Wittgenstein's scare quotes and take this observation as our point of departure. I think it will lead us fairly directly to what Wittgenstein can actually prove. Once we know where this line of thought seems to be going, we will look at his argument in more detail.

Suppose, for example, that I am given a mathematical rule. According to Wittgenstein's own observation quoted above, I can attribute some semblance of "fictional" meaning to it in the privacy of my own head, and then take what I consider to be the appropriate course of action. But things are quite different as to the *justification* of that meaning and the corresponding course of action. If I wish to justify a course of action, I must stop and, as it were, tell myself: "You should do this and not that, because...."

The key word, here, is "should." To say to myself that I *should* interpret the rule in a certain manner for one reason or another in effect requires that I transmit my own will to myself through language. Now bring in Brouwer: Will-transmission through language is uncertain. If this is the case, then it follows that I cannot justify that I should do this and not that. (But note *Philosophical Investigations*, paragraph 289: "[T]o use a word without justification does not mean to use it without right.")

So it does not seem problematic whether I can attribute some meaning to the rule in the privacy of my mind. My actions may not be internally justifiable, but that does not mean that my only choice is blindly to follow the community's practices. If anything, I have too many choices—this is the basic lemma of Wittgenstein's "private language argument," which I discuss below—and it is the justification of a particular choice that presents me with a problem. It is in *that* sense that I follow the rule "blindly": I make a hypothetical judgment about it.

From this point of view, Wittgenstein's argument establishes only the following theorem: *Because* individual will is impenetrable to language, the criteria for *justifying* any course of action are always public.

As far as I know, Brouwer never made that conclusion. He did say that an individual uses language in solitude *only* because "in his thinking science and organization of society *have* to be taken into account."[6] It is not particularly clear what Brouwer means when he claims that something happens *only* because it *has* to happen. Wittgenstein made it abundantly clear: Whatever I may do in my solitude, I have to step outside of it if I want to justify it.

This, however, seems to be little more than a reformulation of one of the fundamental principles of romanticist language philosophy. Every act of understanding includes a mixture of generality and individuality. That prevents me from being able to claim internal knowledge, because knowledge involves justification and hence outward criteria. But it works in the other direction, too. An individual contribution—an element of "interpretation"—prevents the act of understanding from being justified in a "general" manner. (Wittgenstein does not employ the word "interpretation" on technical grounds; it means something very specific in mathematical logic. I will not adhere to this convention.)

Thus, every event of understanding includes an irremovable component of misunderstanding, which, recall from chapter 4, was a claim made by Friedrich Schlegel some time around 1800 (and elaborated by Humboldt later). Wittgenstein's paragraph 526 addresses the issue in similar terms: "What does it mean to understand a picture, a drawing? Here too there is understanding and failure to understand." This seems to be how Schleiermacher and Humboldt viewed the matter.

Individual interpretations are unjustifiable, and they must rely on outside criteria. In these circumstances, the best I can do in terms of justification is to invoke some pragmatic criteria and ask: Do I interpret this rule (text, convention) as others do? What criteria are there that could help me understand? What "techniques" should I learn? But that does not mean that anything on the outside (language, culture, tradition, nature) *determines* my understanding—it informs understanding by *removing* some modes of interpretation.

To sum up, when Wittgenstein argues that "inner activity stands in need of outer criteria," it seems safe to say that—apart from the important fact that he brings some mathematics to bear on the problem—we are talking about a 200-year-old hat. In one form or another, with great terminological variations, this much was known already to Fichte, one of whose principal insights can be summed up as follows: Without the not-I there can be no I. In fact, it appears that the structure of Wittgenstein's argument vaguely resembles some notions of Fichte's.

But caution is always advised, and it seems prudent to keep in mind that Wittgenstein's intentions may have been altogether different, perhaps closer to Nietzsche's attack on the privacy of less than superhuman individuals. It is quite possible, and it appears to be the prevalent view, that Wittgenstein attempted a "reduction to the absurd" of romanticism in general and of Brouwer in particular. However, if this is the case, then his argumentation must invoke some additional assumptions to which other people need not subscribe at all. We will look into this possibility, too.

Let me bring this rather abstract introduction to a close and turn to (my reconstruction of) the argument itself. Initially, and in an appropriately Wittgensteinian spirit, we will consider a simple example and try to refine it as we go along.

9.1. Interpolating the Self

Consider the following story, taken from a fictional psychiatrist's casebook.[7]

An unemployed mathematical subject, Sue, is to be tested as part of a job application. One of the questions on the test is, "What is the next number in the sequence 2, 4, 6, . . . ?" The administrator of the test, the human resources manager named Hank, is convinced that the correct answer is 8. For Hank, "clearly," the three numbers mark the beginning of the sequence of even positive integers, which is the sequence of values of the function $F(n) = 2n$.

The numbers in the sequence are making their appearance on this test precisely due to the fact they are collected according to a certain rule. Sue must grasp the rule in order to give the correct answer. Her grasp of it is confirmed by her being able to produce the numbers as the rule demands, that is, by the ability to continue the sequence according to this mathematical rule. Sue answers the remaining 159 questions in what Hank thinks is the correct way, but her answer to the above question was, say, 10.

Since she scored highest among all the job applicants, Sue is called in for an interview but is required to explain this oddity. She replies that the numbers 2, 4, 6, are the first three elements of the sequence of numbers collected according to the rule

$$G(n) = \frac{1}{3}n^3 - 2n^2 + \frac{17}{3}n - 2$$

Hank computes a few things on his calculator and confirms that $G(1) = 2$, $G(2) = 4$, $G(3) = 6$, and $G(4) = 10$. So this rule fits the given pattern, as far as the first three instances of it are concerned. Based on the accessible information about the rule—the first three instances of its application—there is no reason to believe that the we are dealing with rule F rather than rule G, even though G gives the "bizarre" answer 10.

Hank is temporarily confused. Carried away by this small victory, Sue argues not only that 10 is just as reasonable a reply as the "expected" 8, but also that, say, 13 is equally justifiable. Based on the accessible information, Sue playfully maintains that the rule she was supposed to grasp could have been thought to be

$$H(n) = \frac{5}{6}n^3 - 5n^2 + \frac{67}{6}n - 5$$

Again, $H(1) = 2$, $H(2) = 4$, $H(3) = 6$, but $H(4) = 13$. And for any integer whatsoever, call it m, there will be an appropriate "rule," or function, such that the first three values that this rule produces are 2, 4, 6 whereas the next one is precisely m. Sue can therefore argue that *any* answer to the question could be thought to be derived according to some rule, and that consequently no answer to the question is somehow privileged in terms of justification.

The lateral thinking Sue exhibits would hardly be an asset for the job of a teller, for which she was overqualified in the first place. Hank thanks her, rejects her as unsuitable, and writes "not a team player" in her file, thereby making sure that she never gets a job anywhere in the banking sector.

Dejected and still unemployed, Sue regrets her playfulness and in a belated fit of conformism begins to believe that there must be an intrinsic justification for Hank's preferred answer. Psychologists who constructed the test must have had a reason to think that there is something in her mind that would make that justification accessible to her. She thinks about it a while, but soon begins to doubt her ability to justify that she can follow any rule whatsoever. Her confidence level became so low that she wondered whether she would even be able to grasp the rule for multiplication by 2. Reconstruct her argument as follows.

Sue may assume that her mind, or that part of her mind that is accessible to her in the moment of justification, has actually operated on only finitely many integers. For the sake of simplicity, we can imagine that those were, say, integers from 1 to 10. The rule F for multiplication of any integer by 2 applies, potentially, to infinitely many integers, or at least to an unbounded number of them. This means that when she is given an integer n, Sue must—in order to justify her belief that she understands the rule—produce the result by following the rule F and thus arrive at the answer $2n$. But in the past she had only applied that rule, and any other rule for that matter, to integers from 1 to 10.

Suppose, for example, that she tries applying it to 11. Sue does the computations and concludes that the answer is 22. She believes that this is true. She believes that she has followed the rule. But can she justify that, based on what she has access to? She doubts *that*. Here is why.

Perhaps, in the past, when she thought that she was applying the rule F, "multiplication by 2," she was actually applying a different one, for example,

$$W(n) = \begin{cases} 2n & \text{if } n \leq 10 \\ w & \text{if } n > 10 \end{cases}$$

Here w denotes any integer whatsoever. The two functions, "rules" F and W, seem identical in terms of what *her* mind contains about them (just as the two rules F and G above were identical in terms of what the text of the test question contained about them). Based on what is "in" her mind, Sue cannot justify the assertion that she had been applying F and not W in the past. Justification cannot rely on beliefs, only on evidence and argumentation. But the traces that her past following of the rule F left in her mind are *exactly* the same as the traces that following the rule W would have left.

Perhaps, then, she had been applying the rule W all along? In that case the result of her following the rule should have been w, not 22. So now it appears that *any* number must be regarded as being equally well justified as the result of the multiplication of 2 by 11; equally well as the one she believes is correct.

That seems preposterous. Surely Sue knows how to count. She could take 11 coins, align them on the table, count them from left to right, stop, go back to the leftmost coin, and continue counting until she reaches the rightmost coin again. She would get the answer 22. However, this invokes *another* rule, namely, the counting rule.

But Sue had only operated on integers up to 10 as her "input." This means, in particular, that she never counted more than 10 things. Thus, the rule for counting is subject to the same challenge. Is there anything in her mind that would justify that the counting rule she had been using to produce the sequence 1, 2, 3, . . . , 10, and that would produce the number 11 as the next one, was not *in fact* some other rule, one that produces the sequence 1, 2, 3,

..., 10, followed by 1? These two rules would have left exactly the same traces in her mind, so she would not be able to say, not without invoking yet another rule, that she was in fact following the counting rule.

This process could be continued indefinitely. But every time another rule is introduced, Sue will be able to evoke only a finite number of instances of applying that rule. At that point analogous objections will apply. Sue thus comes to the conclusion that following a rule can be justified only by invoking a meta-rule on how to follow the rule, and then a meta-meta rule on how to follow the meta-rule, and so on. If we are to avoid an infinite regress of ever higher rules, this must stop at some point. Then this ultimate rule can be challenged, just like the original one was.

There is, of course, nothing special about having seen only 10 integers or about the particular rule of multiplication by 2. There seems to be a finite limit to what we have experienced at any moment in life. Even if someone disputed that, it is quite obvious that there is a finite limit to what we can *say* we experienced. (Speaking and thinking take time, and we do not live forever.)

Hence, the argument can be generalized. Based on what is accessible to me by means of inward reflection, any outcome of following the rule seems equally justifiable. But this means that *no* outcome of following a rule can be justified by invoking internal data. And so we have discovered Wittgenstein's skeptical paradox: "[T]his was our paradox: no course of action could be determined by a rule, because every course of action can made out to accord with the rule."[8]

Let me point out once more how this relates to Brouwer's idea that linguistic rules (even mathematical rules) cannot ensure certainty of will-transmission. One might say, for example, that the psychological test in the above example is intended to transmit the will of the psychologists. They constructed their test believing that the sequence "2, 4, 6, ..." should be continued according to the rule for multiplication by 2. The tested subject Sue, even though she realized that she could have continued the sequence in that manner, still has no way of logically ascertaining that this was indeed the will of the psychologists. She realizes that it could be so, she even seems to be believe that it is so, but she cannot justify why it *should* be so.

She can reinterpret the data that are given to her linguistically in many ways, each of which is equally justifiable based on what is given. Furthermore, even when she decided that *this* is the rule that she was supposed to follow—and thus tried to see whether the rule of her own choice can logically determine her course of action—she was unable to justify that her actions indeed followed *that* rule. Hence, she cannot transmit her own will to herself with certainty, if she takes a detour by way of language.

The discussion also relates to that other problem, the problem of "private language." The argument seems to imply that one person, on her own, cannot claim to have a determinate semantics based only on the knowledge of the syntax, that is, her knowledge of the rule.

The psychiatrist therefore suggested that Sue program the multiplication rule F on her computer. Should an anxiety attack occur, she could apply what she thinks is the rule F and check her results against that of the machine. That seemed to work for a while, but she soon went into a bizarre relapse. She developed her suspicions when she observed that the computer could have only "seen" a finite number of instances of the rule F.

In a series of brilliant and rather expensive sessions, the psychiatrist argued that the rule is in a sense physically realized in the machine itself, so the computer—or the appropriate program, which is stored in its memory bank, thus having a physical quality—actually embodies the rule. He expounds a variant of the functionalist stance: He equates the subject's semantics with the syntax of a PC.

At this point the notes on Sue abruptly end, so we can only assume that she could not pay the bills. The psychiatrist triumphantly observes that she was cured and goes on to the next case. Yet we seem to be left with several questions. How is meaning possible at all? Why do I believe that I am following a rule, if I do not seem to be able to justify that this is indeed what I am doing? How do I make the choice that I make? Is the functionalist shrink's resolution of Wittgenstein's paradox any solution at all?

9.2. Language Games

Let me sum up the preceding section. Since my mind is clearly of finite capacity to formulate anything linguistically, I can only invoke finitely many traces of applying any rule when I am asked to justify my following of the rule. Wittgenstein argues that there is therefore nothing that I can recall from within my own mind that would ascertain that I am following *this* rule and not some other rule that just happens to be the same—in the finite number of cases of which I have produced a record—as the one I think I am following. Since any finite amount of evidence can be made out to accord with a rule whose next instance is anything I want, it seems that I can justify anything I want. Hence, I cannot justify anything. In particular, I cannot justify the claim that I know the meaning of the rule.

How, then, do we get to fix the meaning of our statements? First consider the possibility we left off in the preceding section: the functionalist account.

For the functionalist, semantics is equated with syntax. Following a rule is "ensured" by possessing the finitely many instructions on how to follow the rule. We could, for instance, program a computer to follow the rule and declare that this computer now embodies the rule itself. But suppose I program a computer to perform multiplication by 2. Even disregarding the obvious objections, such as the fact that there could be many different programs that "embody" the same rule, the possibility of malfunction, and so on, the crucial problem remains: If I write a program that will follow the rule, how would *I* know that this is just what it does? This is important, since my intention—if I am a functionalist—is to fix the meaning of my assertions by declaring that they tend to be the same as those of the computer.

If I believe that 3 times 2 is 6, but my program claims that it is 5, would I change my mind? I suppose not. I would assume that it is a simple bug. So I would "debug" the program before I declare it to be the official oracle of my semantics. But to debug it involves testing it. I would thus try to make certain that the program runs "correctly" for the instances of the rule it embodies—instances that are accessible to *me*. I would test its action on the first 10 positive integers, which is all I know.

This is not much better than what we had before. If the computer gives me the answer that 2 times 11 is 22, what justification do I have for that claim? My skepticism regarding my own answer, 22, was not inspired by my thinking that I had made a mistake, or that I was in some sense wrong. On the contrary, I was convinced, just like Sue was in the preceding section, that I was right. I simply did not know *why* I was supposed to be right. The same question applies here. I may believe that the computer actually embodies the rule, that its answers are infallibly correct, but I still do not have any justification of that belief.

Thus, the functionalist account seems somewhat unsatisfactory. It involves making a blind leap into declaring the meaning to be identifiable with the actions of a syntactic machine that, for all its perfection and implicitly assumed infallibility, fails to provide anything like the solution of the problem. In fact, if you turn things around and ask how would *the machine* be able to justify its alleged semantics, the same argument would show that it never could.[9] Let us see how Wittgenstein proposed to resolve his paradox.

Wittgenstein considers the human being as a member of a community. Since it appears to be impossible—according to his "paradox"—to ascertain the meaning of my assertions based on what I have in my head, I could try to perform "experiments" within a community. For example, suppose that I think I know the rule for multiplication by 2. I cannot find anything within myself that would justify the way I act when I think that I am following that rule. I could, however, assert things such as "2 times 11 is 22" and expect to be corrected or reinforced by the other members of the community. I am thus engaging in a *language game*.

The community follows its own conventions on the use of the terms I play with. I can expect that they would reinforce or discourage my use, until I am certified as a competent player—or "certified" as incompetent, crazy, or worse—based on my behavior and my circumstances. One might say that this in some way depicts the notion of "training," "education," or "culturalization" (although the picture leaves one with the uneasy impression of a society as a boot camp based on B.F. Skinner's doctrine of operant conditioning).

Once I am certified as competent, it will be assumed that my meaning accords with the accepted one. No certainty is achievable this way, nor is it sought. Even if I am accepted into the community based on my competency, there is nothing that would guarantee that the meaning is now fixed—neither mine *nor* the community's. For instance, I could exhibit behavior that would make the community revoke my licence as a player (e.g., if I am caught smoking in my office, right under the sign that reads "this is a clean-air campus"). Conversely, I could employ some previously acceptable term without knowing that it will now cause my neighbors to shun me.

Let us tread carefully over this territory: We are approaching an ideological minefield. There are two rather different conclusions one might draw about the relationship of the individual and the community.

One could say, based on what the argumentation demonstrates, that the community, the collective, is involved in motivating my interpretive acts but that it does not necessarily supersede my conviction. The community of players of a particular language game guides the interpretation of rules. This is natural and in some sense obvious. It would be strange to say that culture, education, tradition, community, or my experiences of the physical world have no bearing whatsoever on my interpretive practices. It is also fair to admit that I am indeed "trained" and indoctrinated in various ways. But it *also* follows from the above argument that even upon extensive training individuals can *always* challenge the grounds of justification of any given rule, as we saw in the case of Sue above.

Nonetheless, it is possible to steer the conclusions from the "private language argument" in a completely different direction. If I believe that meaning anything by anything involves my being able to justify it—or if I happen to be one of the people who believe that meaning resides solely "in" justification—then it seems to follow that I cannot have any semantics of my own. To have any semantics whatsoever, I must follow a cultural convention. These conventions are drilled into me daily by my culture, a tradition into which I enter upon birth.[10]

Putting it somewhat crudely, the community programs me and debugs me during the language game that is my life. Conversely, I use the community just like the functionalist shrink suggested Sue should use a PC: I identify my meaning with what *it* does. It therefore appears that in this case we have a kind of functionalism on our hands. Words perform a certain function ("use") in the system of cultural conventions. I am trained to observe these conventions; the only way in which I can escape them is by making a mistake, by unwittingly causing some "infraction" of the rules. These infractions are what I mistakenly attribute to my own "creativity."[11]

So there are two very different ways of reading Wittgenstein. The ambiguity, I think, stems from the serious internal tensions of Wittgenstein's attempt to temper Brouwer's more eccentric exclamations, to disentangle the vague and confusing terms such as willing, self, and interpretation and yet at the same time keep the natural wealth of ordinary language. For Wittgenstein, meaning is sometimes akin to fiction, and at other times it involves justification. Understanding is in one place akin to art, while in other places seems closer to "formal" linguistic justification.

In short, it would seem that Wittgenstein attempted a compromise between formalist and romanticist viewpoints. The ambivalence is best reflected in his description of mathematics: "Of course, in one sense mathematics is a branch of knowledge,—but still it is also an *activity*."[12] On the one hand, he defends Brouwer's notions about the uncertainty of language. On the other hand, he toys with the formalist idea that meaning is ultimately in justification. Let us look briefly at this "formalist" side of Wittgenstein.

I noted in chapter 6 how the very concept of language games seems in part inspired by a view rightly or wrongly attributed to Hilbert: Mathematics consists of "empty formula games" whose only meaning resides in the act of demonstrating something according to certain rules. Since Wittgenstein wrote *Philosophical Investigations* after formalism's failure to find a single "metagame" that justifies all other mathematical "games," it seems almost natural that he would embrace the fragmentation of language games into possibly incommensurable communities and "forms of life." In this sense, language *itself* mimics the old romanticist idea about the uncertainty of communication and relativization of truth to linguistic-historical communities. (These communities may in fact be speaking about the same truth, but they cannot justify that they do, due to communication problems.)

Thus, it becomes unnecessary—for Wittgenstein—to invoke the will, interpretation, individual consciousness, the ineffable inner flux, and similar "metaphysical" notions. They all "divide out." All Wittgenstein has to do to support his viewpoint is to take romanticist argumentation and *re*formulate it in terms of language *itself*. This method, however, inscribes the entire romantic thought so deeply into Wittgenstein's arguments that it is not clear if he can ultimately establish anything substantially new, other than by decree.

For example, with regard to following a rule, Wittgenstein says that it is not possible to do it privately (para. 202) because grasping a rule requires simultaneously "obeying the rule" and "going against it." Quite similarly, for Fichte, the I can be sensed only if it has something to "go against," the material world, language, culture, or some larger and more inert not-I that *resists* my creativity and my will.

One might further compare Wittgenstein's statements about understanding language being akin to understanding a piece of music, and assertions such as "grammar [. . .] only describes and in no way explains the use of signs," with Schleiermacher's notion that linguistic rules do not absolutely determine their own application and that individually attributed meaning is always reestablished in social practice.[13]

From this point of view, it would appear that Wittgenstein's stance can be summed up as follows: I am I *and* my circumstances, but I can only *observe* my circumstances, so the I, or some deeper self, the will, ultimately "divides out" of the language game. To be even more precise, it divides out from the language games construed on the model of *object and name* (para. 293). The self, the will, or whatever else we might *name* it, is not an *object*.

Thus, it seems that Wittgenstein inferred from Brouwer's statement that "all speaking and reasoning are an attention at a great distance from the Self"—or from romanticist ideas about the self as a continuous action, as something that is not some*thing*, not an "it"—that the self and its mysterious inner activities should drop out of a good deal of language games. They should be replaced by language and the self's circumstances as objects of study. (This is exactly what Foucault proposes.)

What, then, *is* it that gives meaning to statements? It is not cultural conventions alone. This would mean that semantics is in the culture syntax. That would be a form of functionalism, which fails by Wittgenstein's own argumentation. The "private language argument" can be generalized. The same argument that requires *me* to have "outward criteria" for (the justification of) my semantics would also require the formal culture-structure to have some outward criteria. What could those be? One might say that meaning arises in part from some "unidentifiable subjective contributions," from individual activities that are outside of language and conventions.

However, there is little room for the romanticist flux of creative activity in Wittgenstein's language game: "It" is not an object, it cannot be named, and hence no justification can invoke it. It cancels out. So how *do* things get to mean anything? It almost looks as if Wittgenstein would like to invoke something like Heidegger's "mathematical," without, however, mentioning it at all.

This may or may not be the case, but it would explain a few things.[14] Heidegger was of the opinion that subjectivity, "psychologism," romantic individualism, needs to be "destroyed" in some sense. Wittgenstein argues the same, but on pragmatic grounds: Whatever it is that is specific to each individual, it is not accessible to language, and we might as well ignore it. We must make the "linguistic turn" and study language itself.

That was also Heidegger's project (which in fact derives from Nietzsche): to substitute the study of language *itself* for the philosophical paradigm of individual consciousness. The basic idea is that language analysis would reveal the underlying conditions of our understanding, the "prejudices" that underscore all reasoning.

This is, of course, mere speculation, and should be taken as such. But with or without Heidegger, Wittgenstein's strange concoction of "intuitionist romanticism" and "ordinary language formalism" is something to keep in mind. It appears that just such an ambivalence, the specter of Wittgenstein's split personality, as it were, haunts and informs a good deal of postmodern thought.

9.3. Thermostats "Я" Us

Alan Turing, Wittgenstein's Cambridge colleague, was involved in philosophical considerations about intelligent machinery. Turing took a philosophy of mathematics class from Wittgenstein. That was in 1930, while Wittgenstein was developing some of his ideas about Brouwer's claims on the uncertainty of will-transmission. Turing was also in some way interested in intuitionist notions, in the "constructive" approach to mathematics, and, as I showed earlier, in the relationship of the continuum and the limits of theoretical computability. What we examine briefly here is that Turing's ideas about testing machine intelligence seem to bear some resemblance to Wittgenstein's views.

Turing asks the question, How would we know whether a machine is intelligent? After a long discussion he proposes the following criterion, known as the Turing test. If I *interact* with something and I think that it exhibits intelligent behavior, then I may consider it intelligent.

What Turing seems to be saying is that we must get involved in a "language game" with the machine, and that the way in which the machine may be accepted into the community of "competent"—that is, intelligent—players, should be the same as the way in which people are accepted into the community of intelligent people. Ultimately, I have to observe the behavior of the thing, its "functioning," in order to judge its intelligence. There is no other way. There is nothing in the syntactic rules that govern its behavior that allows me to make such a judgment *a priori*. I cannot a priori say that such and such a program *will* exhibit intelligent behavior.

We have seen part of this reasoning before. I may write a computer program that supposedly follows a rule, but how do I know that it *does* follow that rule? I can only judge that by comparing its behavior with what I consider to be the appropriate way of following the rule. It does not help that I wrote the program.

Let me "unpack" this further, and put it in the context of Turing's test:

- There is no *formal* justification—or definition—of intelligence.
- The only relevant access we have to the "mental states" of a machine is our observation of its linguistic behavior.
- So it is only through an interaction with a machine that we can come to judge its competence at following a particular (intelligence-indicating) behavioral pattern.

- If the community of humans has criteria for judging such competence (of humans), the criteria should be applied to judge the competence of the machine.

If we now replace "machine" by "person," "intelligence" by "meaning," and "interaction" by "language game," we get a nice structural similarity to Wittgenstein's solution of his paradox:

- There is no *formal* justification of meaning.
- The only relevant access we have to the "mental states" of a person is our observation of the person's behavior (and a variety of "circumstances").
- So it is only through a language game that we can come to judge a person's competence at following a particular (meaning-indicating) behavioral pattern.
- If the community has criteria for judging such competence, the criteria should be applied to judge the competence of the person.[15]

The similarity goes a little further than that. Turing, like Wittgenstein, showed concern for the problem of will. For Wittgenstein, structural generalities, reasoning, "rules" cannot determine any course of action. Not even I myself can a priori *know* my will, in the sense of being able to justify my actions internally. That does not mean that there is no such thing as will, only that it eludes language, and therefore justification. It "divides out" of language games, but it would be funny to claim that something "divides out" if it was not there to begin with.

Turing, for his part, distinguishes carefully between *discipline*, whose conceptualization is represented by the calculating machine, and *initiative*—the "residue" of human intelligence, its fundamental constituent that has not been considered so far, and that we should strive to copy (i.e., *simulate*) in machines.[16] But if there were something that one could definitely say about this undisciplined "residue," it would instantly be brought up to the level of logical description and thus be subject to discipline, that is, machine modeling.

This is, of course, the task that Turing sets for himself. Valuable lessons might be learned along the way. However, if the project could be completed—in the sense of "proving" that some machine is intelligent—Turing's test would be utterly redundant. If intelligence were something static and immutable, something ultimately subject to complete logical description, then I could simply check whether a machine (or a program) satisfies the a priori syntactic conditions for being intelligent, and that would be it. Turing is more cautious.

It is here, perhaps, that the philosophy class he took from Wittgenstein shows its influence: How would I verify that some program follows these intelligence rules? I cannot be certain a priori that *I* am following the rule for multiplication by 2. I can only make such judgments based on my beliefs or on the available public criteria for competence in following the rule, which are in no way fixed. So there can be no "proof" that the machine is intelligent. Intelligence, like meaning, is something that always has to reestablish itself in social practice.

The Wittgensteinian spirit of Turing's test is typically ignored. As a consequence, Turing's alleged followers emit a continuous stream of announcements to the effect that we will soon exhaust all the items on the intelligence checklist and thus construct human intelligence in a machine. But how would we know that we exhausted *all* such items if we do not already know all of them beforehand? How could we be certain that no hitherto undisciplined residue of logical description will ever surface? In other words, how would we *know* that the machine *is* intelligent? Apparently, that is not a problem. Computer scientists will be there to tell us that it is. To quote Wittgenstein: "A man says 'I know how tall I am' and puts his hand on his head to prove it."[17]

The possibility of artificial intelligence has been disputed by several prominent scientists and philosophers. Let us look at one of the arguments that has been the source of some controversy: the "Chinese Room" argument.

The Chinese Room is a thought-experiment invented by the American philosopher John Searle: Imagine that I am in a room that has pieces of paper with Chinese characters written on them, and a book of instructions that tells me how to proceed when somebody slides some pieces of paper with Chinese characters on them under the door. To the outside observer, if I follow the rules in the book, and if the rules in the book are some appropriately detailed version of the Chinese grammar, it would appear that I am able to respond to Chinese language *as if* I understood it. But, in fact, I do not understand the first thing about it. Since I *in fact* do not understand what I am doing, I know that I am not behaving intelligently—even though it may appear to others that I am.

Now replace "room" with "computer," "book" by a computer program, and me by the central processor. It follows that even though I might be perceived to be behaving as if I understood something, I have no idea what I am doing. By analogy, if *I* do not have the foggiest notion about what I am doing, then the computer cannot have a clue either. It cannot *be* intelligent, though it might behave as if it were. So we can at best only simulate intelligent behavior on a computer.[18]

The conclusion is, I believe, correct. But what seems to me a little unpleasant about this argument is Searle's claim that I cannot attribute any meaning whatsoever to what I am doing in the Chinese Room. In that case there appears to be little left to prove, since I assumed even more than I wanted to prove, namely, that *I* have no semantics in the Chinese Room.

I would like to consider this in more detail, so let me translate the Chinese Room experiment back into its original form: Wittgenstein's "private language argument."

Suppose you give me a booklet with instructions on how to follow the rule for multiplication by 2, and send me to a desert island called Chinese Room. Once I am there, I keep receiving messages in bottles. The messages ask me to perform more multiplications by 2 and to send out my responses. I apply my knowledge of the multiplication rule, but how can I be sure that I am indeed following the rule? According to Wittgenstein's argument, I cannot justify my responses based on anything that is available to me. That does not entail that what I do—in the Chinese Room or on the desert island by the same name—has no meaning *for me*, although I cannot *justify* my attribution of that meaning to my actions.

However, this is not the case according to Searle. For him, even my innermost *intentions* must be codified as part of some cultural convention. Indeed, Searle's theory of language (speech-act theory) classifies the types of individual intentions beforehand. It is a meta-theory that governs all possible speech-acts. According to this view, I can mean something *only if* I justify it by reference to a certain meta-theoretical convention. So meaning resides *in* justification.

We are already familiar with this formalistic axiom, and we know that it can lead to assertions such as Foucault's "there is no author." In Searle's case, the same idea surfaces as the claim that I cannot attribute any meaning to anything during my solitary confinement to the Chinese Room.[19] But this is not entirely beyond doubt: Wittgenstein himself describes a situation in which he can attribute a "fictional meaning" to an unknown symbol. (E.g., I could imagine that what I am doing in the Chinese Room is keeping the books of an eccentric Hong Kong mobster.)

Hence, barring further ideological assumptions about the divine omnipotence of cultural conventions, the Chinese Room experiment demonstrates *only* what Wittgenstein's "private language argument" demonstrates. I (and the computer) may have some semblance of semantics in complete solitude, but I (and the computer) cannot justify it. This still leaves open the possibility that the computer could in fact be intelligent. I find this unsatisfactory, so I would like to try a different approach.

The problem seems to be that, seen through the formalistic lens of language alone, the entire phenomenon of consciousness reduces to what it can *say* about itself. Language is certainly a necessary and a crucially important aspect of human existence, but this viewpoint *identifies* self-consciousness with what we see of ourselves in the mirror of language (thus equating consciousness with cognition: knowledge, justification, etc.).

Perhaps we can learn something by unlearning this idea. That appears to be what is needed, since the artificial intelligence thesis, in its strongest form, confuses consciousness and cognition to the extent that it claims that a computer can *be* a self-conscious being simply because it satisfies some strictly linguistic criteria. Let me first address the idea that a computer can become self-conscious, and then we will see how this relates to its intelligence.

Suppose that for some reason, perhaps out of morbid curiosity, I decide to buy a computer that can allegedly become a self-conscious and intelligent being. I bring it home and flip the switch. Programs are now loading up, the machine is coming into consciousness about itself, and, after a while, it supposedly becomes self-conscious. A miracle has occurred.

Let us be cautious and replay this event in slow motion.

In front of me there is a computer. For the time being, it is completely unconscious. In particular, it has no consciousness of itself, no awareness whatsoever, not even a dimmest trace of self-acquaintance. It is an expensive *tabula rasa*. Then I switch it on. The machine, at this point obviously still clueless about itself, performs certain formal-logical operations, and thus becomes conscious of—itself. But how could it possibly arrive at this insight, if it had no familiarity with itself to begin with? The machine has no criterion at its disposal that would allow it to identify anything as being *itself*. It can recognize something *as* itself *only if* it already had some awareness of itself. Since it did not have such an awareness to begin with, it seems unlikely that it will have any grounds to become self-conscious.

This is in fact a reformulation of the vicious circle that Schelling discovered in Hegel's philosophy, some time in the early 1800s. Hegel's spirit—a logical creature like the Turing machine—starts out with no consciousness of itself, performs some logical operations, and then becomes conscious of itself. Schelling asks: How can the spirit get the idea that it is precisely *itself* that it is now conscious of, if it did not *already* have some awareness of itself? And supposedly it did not. (This is why romanticist thinkers introduced the notion of "immediate self-consciousness," "absolutely transcendent being," and other intimidating stuff: We crucially depend on language, but in the beginning there has to be more than just a sign.)[20]

As for the impossibility of an actually intelligent machine—that is, a machine that *understands*—here is an argument more modern than Schelling's. Can one understand something without being conscious that one is understanding something? Maybe. But that there is understanding without consciousness seems a rather difficult proposition to prove, at least for a conscious being.

So it would appear that there is no understanding that is not also a consciousness of its being an understanding. Understanding presupposes self-consciousness. Since, as I observed above, self-consciousness is not likely to occur in a machine, it follows that understanding cannot occur in a machine—at least on the "usual" understanding of understanding. This argument, in a somewhat different form, has been around at least since Sartre published *Being and Nothingness* (1943).[21]

These arguments have not gone unchallenged, so I do not wish to press the point too much. But whether or not one agrees with Sartre and Schelling, it seems safe to say that their objections cause difficulties for the strong variant of the artificial intelligence thesis. These and other relevant problems are not likely to be resolved by issuing decrees from the Olympian heights of a fancy computer lab.[22]

The weaker version of the artificial intelligence thesis, which states that intelligence can be simulated on a machine, seems rather trivial and I see no reason to deny it. Anything can be simulated, badly. How good a particular simulation may be, that can be judged only in practice. In the case of intelligence, such judgments are based on all available ethical, social, scientific, political, and other criteria. In short, it is a judgment made by people, because people attribute meaning to things.

A computer scientist like John McCarthy is therefore at liberty to believe in his wondrous thinking thermostat. The claim that the thermostat can be seen *as* having "mental states" is not in itself an absurdity. The behavior of a thermostat allows *McCarthy* to attribute such a semantics to this structure. He can see it *as if* it had three mental states: "it's too cold," "it's too hot," and "it's just right." That does not mean that it *has* these states of mind. And therefore Searle—who does indeed have a bit of fun at the expense of McCarthy and his thermostat—can just as consistently describe this fascinating device *as* an unintelligent piece of bimetallic strip.

People seem to be able to hypothesize meanings of things, discard them, or reestablish them in praxis, based on all that is available: syntax, theoretical framework, discipline, methodology, observation, culture, tradition, myth, experience, intuition, generalization, induction, analogy, initiative, invention, creativity, ingenuity, imagination, fantasy, pleasure, sheer guesswork. And we always—to go back to Kant's expression—submit our stories to the jury of other fallible people.

I think we are now ready to take a closer look at the "postmodern condition."

10
POSTMODERN ENIGMAS

I have never said anything like that.
—Jacques Derrida

Now that we have a reasonably large data bank of analogies at our disposal, we can begin a critical reconstruction of the basics of "postmodern" argumentation. Derrida's work is particularly interesting in the present context, because he seems aware of its mathematical tributaries.

I believe that at this stage we are in a position to infer what appears to be one of the "central" insights of the otherwise decentered thought-constellation known as poststructuralism. More precisely, we can derive this corollary by means of Poincaré's critique of impredicative definitions and givenness of identity. That indeed seems to be what Derrida has done (although for the sake of clarity I will suppress the dark lyricism that clouds his writing).

This, of course, does not mean that Derrida uses the trick with the same goal in mind, that he agrees with intuitionists philosophically, or that he lifted his ideas from a couple of mathematicians. It seems that Derrida is set on rejecting intuitionist psychologism, idealism, and (as he says) spiritualism, while simultaneously trying to import intuitionist argumentative *techniques* into linguistic formalism. So there are some differences, and it is good to keep them in mind.

Nevertheless, important features of Derrida's argumentation are clearly understandable from a mathematical viewpoint and, as he himself has hinted, can be placed in a mathematical-historical context. This is very exciting and I emphasize this connection first. In the final section, I briefly address the more popular forms of postmodern thought.

10.1. Unspeakable *Différance*

If there is anything that initially seems to connect poststructuralism and intuitionism, it is the shared intention to demonstrate that (mathematical) "texts" do not carry their own total meaning within them like an elusive "presence" that is merely decoded in the act of understanding. That is a start. We also know that the basic strategy of intuitionist critiques was to argue that mathematics cannot be identified with logic or language. Recall, further, that the case was argued by exhibiting "something" that is beyond language, something that is in a way "the other of language": the continuum, the medium of free becoming.

In what seems to be a similar vein, Derrida himself says in an interview from 1981 that his work "is always deeply concerned with the 'other' of language. I never cease to be surprised by critics who see my work as a declaration that there is nothing beyond language, that we are imprisoned in language; it is, in fact, saying the exact opposite."[1] The opposite, I suppose, would be that there is something beyond language, that we are not imprisoned in it. In that much, at least, Derrida is in agreement with various mathematicians who criticized the reduction of mathematical activity to language or logic.

So we have some common ground, perhaps a little thin and slippery, but a common ground all the same. We should take a closer look at how far it might extend. I will begin by working up to Derrida's "basic theorem" in a manner convenient for comparing it with Poincaré's critique of identity.

Let us return to the case of functionalism ("structuralism") and its account of semantics. Whether we agree with it or not, we can certainly imagine what would happen if the functionalist project fulfilled its formalist dream of embedding the entire human experience within some ultimate syntactic structure, a sort of ultimate *langue* or "discourse" that comprises myths, language-in-general, social relations, the derivatives market, and other things. This ultimate network of networks is what one might call "text-in-general." The text-in-general is functionalism's equivalent of the Book of Nature.

The meaning of this book exists in the mind of God, wherefrom parts of it are occasionally downloaded by the chosen ones. Derrida wants to dispute the whole notion of the ultimate meaning of this ultimate book, which is why he says "text-in-general" rather than "book." I will initially speak of some generic structure, at least until we begin to understand the difference between a book and a text.

Suppose, then, that we have some total structure, the structure of all structures. How is it given to us? Is it simply finite, with, say, three or four billion elements? It does not seem to be. The world as we know it may or may not be infinite, but it at least seems to be "unbounded." New things appear every day; new human beings are popping into existence by combining the finite number of letters of the genetic code in a new way; theorems are proved in the finitary scheme of mathematical proofs; new materials are produced by combinations of the finitary grammar of chemical elements.

Staying with the description of the world as a text of one kind or another, it would seem that new parts of the text are always being created, even though the total number of textual units is finite. Therefore, it would seem that our ultimate structure is not a finished book, a book *already* written, by God's finger as it were, but that it is always in the process of *being* written. Thus, Derrida's expression "the end of the book" means nothing more and nothing less than this: *The book has no end.* This observation permits Derrida—who argues such metaphors with great literary skill—to conclude that "writing," as opposed to reading that which is already written, is a more appropriate concept through which we may envision the ultimate structure.

The structure itself is never given in its totality, nor can it be so given. What we can study of it is its grammar, the rules of its formation, the rules according to which new elements are produced. Thus, our sciences, rather than remaining in the mode of "reading the mind of God," should better be looked upon as a special case of an activity that Derrida calls "grammatology," a science of *writing* in some very general sense of the word.

Later we will see what this science might look like. For now, we can observe that grammatology is concerned with writing, which is active, involves change, difference, creativity; whereas the entire project of Western metaphysics—at least according to Derrida—is supposed to have been concerned with the static, frozen landscape of an ultimate book that has already been written, and so presupposes an immutable "presence" of a meaning that is to be discovered in it. The new science of grammatology should take into account the fact that there is no ultimate meaning present "in" the text. There is only an endless opening, produced by incessant "writing." Meaning is always an opening toward new meanings.

To set himself up for his argument, Derrida invokes a myriad examples and metaphors that, to him, signify the fakery behind the notion that speech—due to the "presence" of the speaker—is to be considered primary to writing, that writing is merely a record of speech, an encoding of something previously present.

Most interesting for our investigation is the fact that he explicitly invokes mathematics as the counterexample to that school of thought. Mathematical "objects" ultimately cannot be *present*ed. In this sense mathematics is always a text, a product of writing and a distancing from speech. We saw that formalism can be interpreted in such a manner: Ideal objects are products of the method ("writing"); they have no meaning prior to that; they are not *re*-presentations of previously present ideas.

Hilbert's formalism is based on a similar reflection. Mathematical notation brings forth new concepts that cannot be held back by "anthropocentric" discussions as to whether the formally introduced objects represent something already intuited or intuitable.

Most mathematicians are familiar with the idea that having the "right" notation is sometimes a crucial part of the process of invention, and that symbolic manipulation may retroactively influence our intuitive understanding of the problem at hand. Here is what Hilbert says in an article from 1928: "To make it a universal requirement that each individual formula be interpretable by itself is by no means reasonable; on the contrary, a theory by its very nature is such that we do not need to fall back on intuition in the midst of some argument."[2]

This was also recognized by Weyl, who advanced the idea that symbolic constructs through which "consciousness jumps over its own shadow" are necessary, but that they nevertheless continue to be supported by an activity peculiar to the human mind. Otherwise, everything becomes an empty game of symbols, as "proposed by the more extreme branches of modern art."

Derrida, for his part, considers such symbolic constructions, "writing" that is not a mere record of a writer's preconceived "story," to be deeply subversive of all metaphysics. He wants to explore the consequences of this idea to the fullest. In an interview he gave to the Bulgarian-born French philosopher and writer Julia Kristeva, this is precisely what he says: "The effective progress of mathematical notation thus goes along with the deconstruction of metaphysics."[3]

So it may seem at first that Derrida is approving of formalism. One is naturally tempted to note that it was Poincaré who had described logicians and formalists as writers who know only grammar but have no stories to tell, which, one imagines, makes them the precursors of grammatology in some way. Poincaré's goal was to show that mathematical texts are nothing without the unifying act of a human being who "breathes life" into the otherwise sterile logical argumentation. Derrida's aim is somewhat different. He does not argue the case of human beings' capability to "breathe life" into formal texts. Instead, he will draw some radical conclusions from importing the notion of writing-in-general into functionalism.

But it would be a mistake, I think, to say that Derrida is a formalist in any straightforward sense. He assumes, let us say, the guise of a formalist, but vehemently denies that his position (whatever it is) is a formalistic one. Indeed, it is quite interesting that he uses precisely the phrase that intuitionism attached to formalism to distance himself from the latter. Recall that mathematical formalism was characterized as an arbitrary game of formulas, of which many people complained that it is not essentially different from a game of chess. Faced with a similar accusation, Derrida said: "People who wish to avoid questioning and discussion present deconstruction as a sort of gratuitous chess game with a combination of signs [. . .]."[4] He is even more explicit in the interview he gave to Kristeva: "We must also be wary of the 'naive' side of formalism and mathematism, one of whose secondary functions in metaphysics, let us not forget, has been to complete and confirm the logocentric theology which they otherwise could contest."[5]

It seems safe to conclude from this that Derrida is in fact after a critique *both* of "idealist" intuitionism and of "naive" formalism. He proceeds by borrowing from both. This unlikely sublation of two completely opposed schools of thought serves its purpose in Derrida's attack on all forms of metaphysics by all available means: "Grammatology must pursue and consolidate whatever in scientific practice has always already begun to exceed logocentric closure."[6] Let us see how he manages to consolidate these apparently incompatible outlooks.

Derrida begins by observing that functionalism cannot ignore the fact that it is dealing with a structure whose nature is ceaselessly to produce new elements of the structure. The structure seems to be given in terms of a finite but *generative* grammar: a few syntactic rules that produce new elements. This is why he speaks of the "text" as opposed to the "book." This generative nature of the structure could cause some difficulties with the "mutability of meaning" of structural units.

Consider the case of arithmetic, which must be part of our "text-in-general." The meaning of structural elements, their identity, on the functionalist account, must work itself out for me through what we called "structural differentiation." The identity of some symbolic numeral is *in* how it relates to other symbolic numerals within the structure of arithmetic (and, possibly, in relation to many other things). If we agree that this is a definition, then it is what Poincaré called an *impredicative* definition: An element is defined by an appeal to the totality to which it belongs.

Now consider the following description: "the smallest integer not nameable in fewer than nineteen syllables." I can "in principle" name each positive integer, and there must be one that is the smallest one that cannot be named in fewer than nineteen syllables. So my description names an integer by in effect talking about all integers. It is an impredicative definition.

But think about what I have done. I have just named an integer that cannot be named in the way I named it. The baptism I performed consists of eighteen syllables, and asserts that the thing I baptized could not have been baptized in fewer than nineteen syllables. This innocent exercise, known as Berry's paradox, is a nice example of the problems involved in impredicative definitions. If I am to think of integers as a fixed totality of immutable things, and of my semantics as being forever fixed and immutable, then I seem to have a paradox.

Yet, not all is lost. It appears that there are (at least) two ways to prevent this trivial discursive item from producing logical inconsistency. I could say that my generative grammar produced an element (the statement above) whose introduction forces me to reset the process of signification and thus change the functionalist semantics of my structure. Many people find this repugnant. But if I insist that I understand integers as an immutable totality, it seems that I can find a way out of my predicament only by admitting that my understanding of what it means to be "nameable" has changed. This is equally repugnant to those who think that only the immutable is excellent.

It seems that something must change, *unless* I invoke some higher principle that would resolve the paradox and ascertain the stability of meaning. Indeed, the standard way of dealing with such situations is to keep the "object-language" (arithmetical stuff) separate from the "meta-language." On this higher level, we can make assertions about the lower level without getting into self-referential paradoxes. So it is only by means of an appeal to some meta-principle that I can resolve the paradox and guarantee the semantic identity of the structural units. This higher principle—which Piaget explicitly invokes in *Structuralism*, in which he argues that structures must be assumed to be "self-regulating"—is what in postmodern jargon would be called "the center."

However, we are dealing not just with arithmetic, but with some ultimate discursive structure of which arithmetic is part. We are dealing with the text-in-general, outside of which no knowledge is possible. (Knowledge involves justification; justification involves language, text, etc., and thus remains within the boundaries of the text-in-general.) We are, so to speak, at the top floor: There is nothing knowable outside the structure, on some wished-for meta-level.

Thus, if our ultimate structure happens to have a center, then there seem to be two possibilities: Either the center is part of the structure, or it is outside of it. If it is outside, then I certainly cannot know about it. If it is inside, then it is itself unguarded from the possible semantic transformations incurred by "writing," and hence, by definition of "center," it ceases to be the center. Therefore, functionalist semantics cannot be guaranteed to have the desired stability, immutability, continuity, and so on. This is what Derrida calls "the undecidability of meaning."

We are now standing on the ground of poststructuralism. It seems to follow from this discussion that there is no such thing as a grounding, central principle or, more precisely, that there is no such thing that is knowable by *us*. So we invent it, assume it, make it up for our own convenience. This amounts to claiming that our knowledge of our allegedly existing semantics is not vulnerable to historicity and "writing." This is presumably what Derrida wanted to say when he wrote the following:

> The concept of a centered structure is in fact the concept of play based on a fundamental ground, a play constituted on the basis of a *fundamental immobility* and a reassuring certitude, *which itself is beyond the reach of play*. And on the basis of this certitude anxiety can be mastered, for anxiety is

invariably the result of a certain mode of being implicated in the game, of being caught by the game, of being as it were at stake in the game from the outset.[7]

Thus we encounter Derrida's concept of a decentered structure. Let me be extremely pedantic about this. Our ultimate structure, as far as Derrida is actually able to demonstrate—as opposed to what he might manage to cloak in disturbing lyricism—may or may not *be* decentered. The point is that we cannot justify, a priori, that it is indeed centered. There is nothing *that we know* that can put a restriction on its possible transformations or, more precisely, the transformations of its appearance to us. And so, given time, the structure—as far as we can tell ahead of time—might change, fluctuate randomly, and even decompose.

We can also see from this argument that it is not the infinity of language ("text"), an infinity that we cannot submit to our finite glance, that causes trouble. On the contrary, as Derrida would say, the problem is that language is finite. It is not too big, but too small—there is something missing from it (namely, the center). Our ultimate "text" is always finite, but is given by a generative grammar, which by virtue of being generative prevents language from having a priori given bounds, which in turn causes the problem.

Let me recapitulate, because we will soon get into the wilder areas of application of the same idea. Functionalist semantics defines the identity of structural units impredicatively, by reference to the totality to which they belong. This totality cannot be "totalized" because "writing" induces change and the appearance of new units. So upon the introduction of a new textual unit, I would in principle have to reset the play of differences among the signifiers in order to infer the new functionalist semantics of the text. Since I cannot be in the (cognizant) possession of the central principle that would guarantee that this transformation would preserve the "old" identity of the "old" structural units, I cannot a priori know that their identity would not change. It may or it may not *actually* change. But I cannot make sure that it *will* remain invariant.

Therefore, the identity that can be assigned to structural units by means of structural differentiation alone is always subject to *possible* changes. It cannot a priori be guaranteed to have achieved its full signification—"complete meaning"—simply because we cannot a priori know the (finite but unbounded) totality through which it achieves its signification.

Here is what Derrida says about this:

[I]t is not because the infiniteness of a field cannot be covered by a finite glance or finite discourse, but because the nature of the field—that is, language, and a finite language—*excludes totalization*. This field is in effect that of play, that is to say, a field of infinite substitutions only because it is finite, that is to say, because instead of being an inexhaustible field, as in the classical hypothesis, instead of being too large, there is something missing from it: a center which arrests and grounds the play of substitutions.[8]

French psychoanalyst Jacques Lacan formulated this idea in a memorable way: "[W]e can say that [. . .] the meaning 'insists' but that none of its elements 'consists' in the signification of which it is *at the moment capable*. [. . .] We are forced, then, to accept the notion of an incessant sliding of the signified under the signifier [. . .]."[9]

The syntactically inferred identity that functionalism assigns to the elements of some structure is in some sense incomplete—Derrida likes to say "undecidable"—and it cannot be predicted whether it might change upon the extension of the "field of play." At best, the identity of an element is observed upon resignification, recognized as being invariant, and even then remains a hypothetical judgment. Identity is thus secondary to the concept of difference. It is not an a priori grounded, self-sufficient concept: Identity is not "present to itself."

Let me make a few comments on this basic theorem of Derrida's before we see where this relatively simple argument can be taken.

For a start, the reasoning I outlined above is nothing more than a "contemporized" presentation of Poincaré's discussion of the effects of the circularity involved in impredicative definitions. We went through this discussion before and noted that according to Poincaré:

1. The identity of an object is not given by the structure that governs the interrelations of such objects.
2. The identity of an object is knowable only if we allow the possibility (the flow of time, "history") that this object could change, and then notice that it did not change, that it in fact remained invariant.
3. Therefore, identity is a concept that can come only as an afterthought to the concept of difference, but can never be logically deduced from it (unless you can observe an object throughout the entire history to make sure it remains invariant).
4. Hence, finally, identity *always* already involves a *hypothetical* judgment.

Poincaré's argument could be viewed as a "deconstruction" of the logic of identity, in a sense not so different from that in which Derrida uses the word "deconstruction." Just as Derrida's, Poincaré's "deconstructive" argument consists in working from the inside, by assuming the guise of a logician more meticulous than any other, in order to uncover the "unsaid" of logic, which in this case amounts to revealing irreducible prelogical assumptions, ungrounded presuppositions on which logic rests, and which are beyond logic itself. It is perhaps closer to a "Freudian" analysis of dreams—logic's dreams, as it were—than to classical argumentation.

The critique of the concept of identity is exemplary of this method. It is not that identity is contradictory or that there is *no* identity; it is simply not given in the way logic supposes it to be given, namely, "for free" and forever. From a strictly logical viewpoint, it seems that identity requires another concept, that of difference, through which it subsists, and even then only as a mere hypothesis. (It is an intuitively motivated hypothesis for Poincaré. Derrida's situation is a bit more complicated on this issue, because he does not want to make an appeal to intuition. But let us leave that aside for the time being.)

The list of corollaries that Derrida extracts from these observations is quite long. For example, in an entertaining confrontation with Searle, he was able to undermine the position of speech-act theory because it employs a formal-logical taxonomy of all possible conventional acts of speech. However, just like geometry *presupposes* the identity of objects governed by the axioms—their invariance under displacement—speech-act theory *presupposes* that the meaning of an utterance remains invariant through as many repetitions as one likes.[10]

Derrida's objection is essentially the same as Poincaré's. He says that the same utterance cannot in principle be guaranteed to mean the same through all of its repetitions. This result is known, somewhat unfortunately, as the "impossibility of iteration."

Let us see whether we can find some more similarities with "intuitionist" ideas. We saw that Derrida claims that we are not imprisoned in language, that deconstruction "is deeply concerned with the 'other' of language." The question is how we would conceive of the "other of language," according to Derrida. Clearly, we cannot know it, but if we cannot even conceive of it, imagine it, or fictionalize it, then the discussion is over. Derrida, for his part, not only conceives of it, but even gives it a name: *différance*.

Suppose that I am embedded in the text-in-general, that is, that what I can know and what I can mean come to me by way of the play of differences among the signifiers in our ultimate structure. It is the starting point of Derrida's argumentation that I am not in a position to "totalize" the entirety of this text. Because the semantic identity of its units is impredicatively defined, it follows that the introduction of new elements, through "writing," *could* change the meaning I think some textual unit may have.

Consider the worst-case scenario, in which the introduction of a new unit of the "text" somewhere, anywhere, *does* induce a change that affects the identity of the particular signifier I am looking at. In other words, I am assuming *for the sake of the argument* that we can identify "writing" with the semantic change that it supposedly induces.

So anything that is written anywhere instantly causes a change in meaning. If I want to imagine how this would affect semantics, I have to imagine what writing-in-general—I believe Derrida calls it "dissemination"—might look like. Now, I can certainly imagine some units that may be "written," and even endless sequences of such textual units. I can do that, for instance, by imagining a rule according to which a sequence of further elements can be produced. I can even imagine a list of possible rules that would generate such sequences according to some generic method of "writing" (e.g., a Turing machine). What I cannot ensure is that I have covered all acts of "writing" in this manner.

Therefore, I must in addition imagine that the sequences I envision as resulting from my list of rules are interspersed with some spontaneous acts of writing: people writing checks in restaurants, random acts of writing by dadaist poets, writing that occurs in chemical reactions or genetic mutations, lightning being written in the sky during a storm, and so forth. What, then, is it that I have imagined, when I envisioned endless sequences interspersed with presently unknowable spontaneous acts?

Well, I seem to have imagined Brouwer's continuum, or at least something *like* Brouwer's continuum: an indivisible entity consisting of choice-sequences. Husserl conceived of it as a duality of "moments"—as did Weyl—and hoped to treat it as an object of knowledge. Derrida, as I showed in chapter 5, notes that this duality inscribed into the continuum can never be brought into a (knowable) unity, that our consciousness of it cannot be justified logically: "[T]here is no subject who is [. . .] the master of *différance*." He concludes that what was previously thought to be the rigid unity of the sign opens itself to a continuum of meanings. This endless opening was called "the medium of free becoming" by Weyl. Derrida describes it in what seem to be similar terms: "the meaning of becoming in general."[11]

Thus, it appears that *différance* is somehow akin to the indivisible flow of Brouwer's continuum, which Derrida, however, does not conceive of in terms of time or in terms of free individual constructions. On the contrary, he almost seems to conceive of it as a *place* of some kind, an abyss that is neither intelligible nor sensible. As a first approximation, we could say that Derrida is reconstructing what Heidegger called "the mathematical." It is *like* the continuum, but it is not a construct of individual consciousness (and it has been interpreted differently throughout history). It is, as Heidegger might say, space in which we always already move.

Derrida speaks of "spacing." This spacing is not discrete, nor is it spacing in the sense of having a position in space. It is always a movement, it never *is*, and in this sense, just like the continuum, it is not knowable as an object. It is not a thing, but rather, to use a Heideggerian mode of expression, a no-thing. Here is what Derrida says: "*Spacing* designates *nothing*, nothing that is, no presence at a distance; it is the index of an irreducible exterior, and at the same time of a *movement*, a displacement [. . .]."[12]

We are now in a position to understand why it is that *différance*, while related to difference, is nonetheless spelled in a peculiar manner, with an "a." *Différance*, Derrida says, denotes at the same time the play of differences among the units of our text-in-general *and* a certain distancing, deferral, a separation of meaning from itself.

This might sound a tad fancy, but it is perhaps not so complicated. Such special effects are a natural consequence of the structure of the continuum. If the semantic identity of some unit can only be conceived of in terms of its "location" in the continuum of possible meanings—due to the changes incurred by generating choice-sequences through rules and spontaneous acts of writing—then this identity is "not present to itself." In Brouwer's continuum, there is no present.

Derrida illustrates this nicely, and in the process apparently reinvents what Brouwer would call the "falling apart of a life-moment." He notes that due to the structural possibility of incessant changes, due to the continuumlike flow of *différance* and the correspondingly *possible* flow of semantic identity, every moment of affirmation, of saying "yes," a priori obliges me to reconfirm that I mean it also in some future situation. Thus, my "yes"—even as I am saying it—becomes a "yes, yes." That is to say, my present "yes" is instantly doubled and split apart by its future. Derrida says, "The second 'yes' will have to re-inaugurate, to re-invent the first one."[13]

This seems to be a very clear and concise explanation. There is something deeply ethical and even familiar about it. We may recall that Poincaré was of the opinion that because immutability is not secured by logical axioms, theorems are in effect always reaffirmed, reinvented, reinaugurated by mathematicians and future generations of mathematicians. This is part of the reason why mathematicians bother with the strange activity of giving new proofs of old theorems. The second "yes"—in this case, the new proof—is not a trivial repetition of the *same*: It is a reinvention, a reconstitution in a different context.

Finally, it seems that Derrida is after importing a sort of dynamic "intuitionist" view into the realm of static structuralist methodology. The continuum, "the mathematical," as it is for Brouwer and Heidegger, is beyond language, but it is not a totally unarticulated darkness. It has *some* structure. To this extent, it can be modeled, conceived, taken into account even though it cannot be known.

What is required, therefore, is not a formalistic abandoning of this "other of language," but rather a different *kind* of thinking about structures, a thinking that would, for example, emphasize the possibility of continuous transformations of meaning.

What Derrida wants, apparently, is not a *de*struction of structuralism as such, but its *re*construction in a manner that would include some equivalent of intuitionist mathematics. The result would be a new structural "science" that he calls grammatology. Let me cite a passage from an interview Derrida gave to Julia Kristeva, where he addresses this point:

> [T]he theme of *différance* is incompatible with the static, synchronic, taxonomic, ahistoric motifs in the concept of *structure*. But [. . .] the production of differences, *différance*, is not astructural: it produces systematic and regulated transformations which are able, at a certain point, to leave room for a structural science. The concept of *différance* even develops the most legitimate principled exigencies of "structuralism."[14]

I think that we are now in a position for me to make a few remarks. First of all, I think it is relatively clear by now that the issue of the continuum appears to be an implicit theme in parts of poststructuralist theory. This, I believe, is the proper context in which the attacks on the logic of identity, binary logic, and so forth, should be viewed. For instance, recall that Deleuze and Guattari struggled with the strange logic of the continuum, and finally came up with some amusing psychoanalytic explanations of the "nonbinary" anti-Oedipal reasoning, a liberating form of thought that they attribute to schizophrenics (in a delirium).

Similarly, Kristeva's later work, starting from *Revolution in Poetic Language* (1974), appears to be concerned with exhibiting something that is beyond language, some unrepresentable dynamic entity that she designates in various ways: the semiotic, the feminine, "genotext." (In contrast to actual linguistic expressions, which she calls "phenotext," genotext is the process that underlies the production of those expressions.)

Kristeva claims to have shown that the meaning of a literary text is also blurred into something like the continuum. She even seems to have discovered in this entity something that does not fit the accepted scientific (i.e., "masculine") description of time. This may be what she understands by "women's time."

All of this, in the final analysis, seems to be a series of poetic variations on the theme of Brouwer's time continuum or Heidegger's "the mathematical" (now called "the semiotic" by Kristeva). Yet this is not entirely true. It may be that these thinkers are trying to convey something similar to the intuitionist critique of logical reductionism. However, most of them do not have intuitionist mathematics at their disposal as an alternative explanatory model.[15]

It is perhaps for this reason that—in an uphill struggle to reinvent metaphors that formalism once banished—postmodern theorists produced a series of brave utterances that make little sense mathematically. That might explain the penchant for strange lyricism that marks a good deal of postmodern writing. (This is in a certain sense inevitable when one theorizes about the "other of language." Even intuitionism experienced serious difficulties with formulating a properly *intuitionist* logic, one that would not be translatable into classical logic in any way. The issue goes deeper than just looking for a different logic.)

In its more extreme forms—as we saw in the case of Foucault—postmodern theory refuses to acknowledge even some limited capability of the human mind to contribute to the Great Text. Rather, it is the formal structure of language that somehow speaks through the individual. But how is our sense of continuity explained in this scheme? Even a philosopher as pragmatic as Peirce noted that the construction of continuity involves an activity of the mind that cannot be reduced to linguistic operations alone.

The general idea is that a discrete linguistic structure somehow constructs the continuum and thus replaces the role of the individual mind. What was previously regarded as a free, spontaneous construct of the human mind is now replaced by "randomness" and "undecidability" of language itself.

There are a few problems with this. When I started the comparison of the continuum and *différance*, I said that we should look at the worst-case scenario, where every act of "writing" actually induces change in meaning. But I am not *necessarily* affected by writing that takes place in France, although I may be affected by it. That is to say, the worst-case scenario does not necessarily happen, unless we invoke Murphy's Law as our central meta-theoretical principle.

Postmodern theory sometimes dreamily invokes what contemporary physics calls "action at a distance" to support this arbitrary assumption. However, if texts are material and there is action at a distance, then every event in the universe changes the meaning of every text. In this manner, the concept of writing is exploded to such an extent that it becomes difficult to imagine anything that is not writing, so the entire universe turns out to be a text.

This does not apply to Derrida, who is more cautious. Indeed, he emphasizes that this "dissemination" of meaning is not *necessary*; it is *possible*. But philosophy must necessarily consider this structural possibility.

In any case, Derrida cannot construct the continuum from language if he at the same time claims, as he does, that language is always finite. He takes *différance* as something that precedes what we might imagine in terms of idealized legible writing: It is the space, so to speak, of incessantly changing semantic possibilities of writing. As with Heidegger's "the mathematical," it precedes formal languages and remains beyond their reach. But Derrida goes even further than Heidegger himself. He criticizes Heidegger for thinking of "the mathematical" as the ultimate *origin*, the origin, in particular, of all concepts.

Recall my thought-experiment from before, the one where I "threw" myself into Brouwer's continuum and concluded that I can only conceive of myself as I *could* be, due to the future-directedness of the continuum. I cannot know what I *am*: I always think of myself as I *could* be. Hence, my *existenz*, as Heidegger might say, is "the issue for care" (or "concern").

It is only through "the mathematical" that I can see anything *as* anything. And this, in particular, is how I see myself: as the kind of being that is in its being concerned with its being. That is Heidegger's famously circular "definition" of the self-conscious subject. Note that it is the consequence of the very structure of the continuum.

From this it follows that even our intuitions—in particular, the intuition of time—are derived from the structure of existential understanding (an apersonal "intuition" of the world, mediated by "the mathematical"). That is what Heidegger says:

> By showing that all sight is grounded primarily in understanding, we have deprived pure intuition of its priority [. . .]. "Intuition" and "thinking" are both derivatives of understanding, and already rather remote ones. Even the phenomenological intuition of essences is grounded in existential understanding [. . .].[16]

With this, Heidegger believes he has dealt a devastating blow to both idealism and realism. He spoke of destroying all of traditional metaphysics. The project is an ambitious one indeed. However, suspicions arise as soon as we ask ourselves how Heidegger could, as he says, "display the birth certificate" of all concepts. Whose authority legitimizes this document? Heidegger's?

Derrida's objection is slightly different. Roughly speaking, he notes that if something is the origin of all concepts, then it is also the origin of the concept of the origin. So Heidegger's argument seems circular. This circularity cannot be removed, except at the price of making some rather wild metaphysical claims, and so Derrida proposes that the very notion of the origin be abandoned: "[T]he origin [. . .] was never constituted except reciprocally by a non-origin [i.e., *différance*], which thus becomes the origin of the origin [. . .]. Which amounts to saying that there is no absolute origin."[17] So things evolve, somehow, but not starting from an "initial point." What does this mean?

It seems that Derrida would like to think of "evolution" as unfolding in something akin to the intuitionist continuum, understood as *différance*, the medium of free becoming in which there are no atoms, no beginning, and no end. In this manner, he can salvage something of Heidegger's argument, by renouncing the idea that there is such a thing as *the* origin.

Applied more generally, the view seems to imply that nothing happens due to a single cause, one that we could isolate with atomic punctuality. (In particular, there is no unique, original cause of things.) If this is what he is trying to say, then Derrida might in fact find that some scientists are partly supportive of his view. For instance, the American geneticist Richard Lewontin's *Biology as Ideology* (1991) contains the germ of the same idea, expressed, of course, with regard to his particular area of expertise:

> Modern biology is characterized by a number of ideological prejudices that shape the form of its explanations and the way its researches are carried out. One of those major prejudices is concerned with the nature of causes. Generally one looks for *the* cause of an effect, or even if there are a num-

ber of causes allowed, one supposes that there is a major cause and others are only subsidiary.[18]

According to Derrida, such "single-minded" prejudices regularly lead to reductionism or metaphysics, whether in the form of the absolutely first cause, of a final cause, or both. Derrida seems to have taken the intuitionist critique to an entirely new level, by renouncing Brouwer's mystical individualism, Husserl's "primordial intuition," and finally even Heidegger's notion of the origin. One wonders what is left.

But let us not too hastily accuse him of total nihilism. It is not necessarily the case, as some of Derrida's followers and critics seem to think, that he denies the notions of identity, of unity, of purposive action, and, ultimately, of self-consciousness.

Let me illustrate this by going back to Derrida's example with the "yes" that turns into a "yes, yes." It is not clear how I would be able to say my second "yes" if I did not have the ability to recognize it *as* having some semblance of semantic identity to the first "yes." In the absence of some unifying ability of the self, however historical and contingent self-consciousness may be, however internally divided and nonpresent to itself, this would simply be impossible. The ethics of my promise, of my "yes," would then either vanish or turn into a Dionysian orgy of saying "Yea and Amen" to everything.

Derrida, however, acknowledges the elementary fact that meaningful communication is possible (although it is always imperfect). That presupposes at least a minimal constancy of semantic units: "Iterability supposes a minimal remainder (as well as a minimum of idealization) in order that the identity of the selfsame be repeatable and identifiable in, through, and even in view of its alteration. For the structure of iteration [. . .] implies *both* identity and difference."[19] This is presumably his way of saying that there *is* identity, but that it is not the *permanent* and immutable identity of which logic and "Western metaphysics" have been dreaming. Indeed, the very concept of continuity, with which Derrida seems to concern himself, includes *both* identity and difference. How else could one conceive of continuous change?

The trick seems to be that Derrida does not take identity and difference as "first principles" from which he would then deduce the continuum and continuity. That is what romanticism did: The ego provides the identity, "outward" criteria provide difference, and their dialectical unity appears as our sense of continuous change, of creative action. Derrida, on the other hand, takes what he can have of difference and identity *in* the continuum. As a consequence, we can know neither *pure* identity nor *pure* difference. They cannot be deduced back from the continuum—or *différance*—any more than pure black and pure white can be deduced from gray. They are always intertwined, never simply present or absent (as Derrida likes to say).

This seems to me very striking indeed. Quite contrary to the somewhat tedious postmodern obsession with discontinuity, Derrida is apparently saying there is *only* the ceaseless movement, the continuum, "spacing," *différance*. What we see of it in terms of observable differences and identities is akin to the signature that one of those sublime objects of contemporary physics might leave in our instruments. For a time, it seems that it is "present." But it was just passing through what science-fiction novels call "our dimension."

Something similar is apparently the case with self-consciousness. We all have some identity, which, by the way, Derrida does not deny: "There are subjects," he says; "this is an incontrovertible fact."[20] But if this is so, then there must be something more to self-consciousness than mere "effects" of language. Part of my identity must *precede* language, because no identity can pop out of linguistic structures alone. An agency able to hypothesize identity is always already involved.

On the other hand, not even the thought of the self, as was known already to Fichte, can have any meaning unless the self can distinguish itself from what it is not. To this extent, even the thought of my own identity always presupposes a system of differences and can become intelligible only in a language. "My presence to myself," says Derrida, "is preceded by language."[21]

The formalist "sign" and the romanticist "act"—to use the metaphor I have used before—always precede one another, but neither of the two comes *first*. Thus, the ultimate "chicken-egg" question dissolves in the haze of the continuum. If there is no initial point, "the beginning," the *point* of departure, then the question does not apply. It does not make sense to ask what happens in the beginning, what comes first, if there is no point that we could with atomic punctuality designate as *the* beginning.

And so identity and difference, nonlinguistic and the linguistic, are somehow intertwined—one always follows the other in a "decentered Cartesian circuit," cleansed of the theological connotations that accompany the ideas of finality and absolute beginning—woven in the continuous tapestry that Derrida calls *différance*. That, on a slightly adventurous reading, may be what he means by saying that "the subject is produced by *différance*."

It is an interesting view. One would certainly like to know how Derrida understands the relevance of his position to science. It would be nice to know more about its relevance to linguistics, where the debate between Chomsky's view of the innateness of language and Piaget's theory of pragmatic language acquisition is still going on in one way or another; to physics, where cosmologists are wrestling the question of the beginning and the end of the universe; and finally to mathematics, to which Derrida occasionally refers and which has apparently inspired some of his thoughts. What sort of mathematics would fit his philosophical vision? What would be Derrida's philosophy of mathematics? He has said little about that. Next to nothing. (Actually, nothing.)

More problematically, Derrida seems to have achieved the exact opposite of initiating what could be an important debate with science. His peculiar style has managed to alienate most scientists and confuse even his admirers to the degree that he himself has described as "grandiose."

In the light of some analogies with intuitionism, and in the light of the fact that Derrida's proclaimed goal is to indicate a different and presumably better way of philosophizing—just as intuitionism once proposed a better way of doing mathematics—one is naturally tempted to consider the fate of Brouwer's "rebellion" from many decades ago as a relevant historical a priori.

A revision of such proportions requires less radicalism and more clarity. Otherwise, despite all the philosophical differences between intuitionism and deconstruction, Hilbert's 1922 assessment of Brouwer's "revolution" could very well apply to Derrida's intuitionism-liberated-from-intuition. The whole thing might end up being "only a repetition [. . .] of an attempted coup that, in its day, was undertaken with more dash, but nevertheless failed completely."[22]

10.2. Dysfunctionalism *Chic*

It is probably unfair to let the penchant for misuse of the terms "deconstruction" and "difference" by a remarkable number of people go completely unmentioned. But it seems downright impossible to address this problem with any seriousness. What might one say about the appearance of the term "deconstruction" in the contexts of fashion design and the ceramic arts?

Indeed, the subtlety of Derrida's attempt to combine formalist notions with intuitionist counterarguments seems lost on the more popular forms of postmodern thought, which are prone to oversimplification and frequently confuse being shocking with being convincing. What results from this trivializing operation?

Simplifying in the extreme, it appears that postmodernism views human beings as formalistic dummies in the generative grammar of various "brands" and other units of the formal culture-structure. We consume those brands and identify with them. I am not an "autonomous," "Cartesian" subject, nor do I cherish any romantic illusions about my irreducible individuality. I identify myself with various units of the culture-structure. But the units of this system are defined impredicatively, so there is no continuity and no identity. I speak English. I drive a Honda. I wear Dockers. I smoke Gitanes. This is who I am, for now.[23]

In this sense the vulgate version of postmodernism seems to be the philosophy of a society based in part on an unbridled materialistic impulse, not unlike the one that Johann von Schiller called "stuff-drive" (*Stofftrieb*). But this is only partially true. Postmodernism would apparently also like to approximate the more positive, utopian aspects of Schiller's vision, for example, the dream of a society entirely engaged in artistic play, governed by "play-drive" (*Spieltrieb*)—a world where every rule is to be regarded as part of a game that can freely be substituted for any other game, consumed, revised, or discarded at will. Acting upon the playful impulses is now considered as a deeply "subversive" activity, not as dreamy musings of a nineteenth-century romantic.

Could it be that postmodernism, despite its occasionally vulgar materialism, represents an attempt to revive of romanticist *Spieltrieb*? I think that the hypothesis should not be discarded immediately. It is true that postmodern culture is brimming with verbose indictments of romantic idealism and other kinds of "metaphysics," in a tenor familiar from the days when the positivist Titanic was still readying itself, full of luxurious promise, for its maiden voyage. But consider some of the terms repeatedly used in the vernacular of postmodern sensibility: irony, play, nostalgia, body, desire, language, culture, self-referentiality, transgression of rules, fragmentation, incredulity toward single or ultimate explanatory frameworks, the preference for the minority over majority, the dream of equating philosophy and poetry, the paranoid fear that we are merely bobbing, senselessly, on the surface of a great, chaotic, undirected ocean of the will (now replaced by the "code" or "hypertext").

Each one of these notions—modulo the variations one might expect in the age when "chaos" can be simulated on a personal computer—may be traced back to the age of romanticism. It would be easy to support this claim by concrete examples, in a case-by-case analysis. However, I will restrict myself to a couple of passing remarks.

There were, for instance, nineteenth-century stories in which city councillors turn into bowls of punch for no apparent reason; there were stories of noses that detach themselves from their "owners" and engage in adventures in such a dismembered state; there were novels about body and sexuality; there were plays so self-referential and fragmented, so intent on transgressing the rules, that they would suddenly "jump outside of themselves" and start discussing how this particular play breaks the rules as opposed to other plays. If the authors of these works lived in an era of television commercials, one could wager that they would have parodied them, too. What all these works have in common—and this, I think, ties them to our postmodern times—is the rejection of the Enlightenment idea that rules are eternal and universally binding.

The intellectual melange characteristic of postmodern culture seems, then, to be a curious simulation of romanticism *and* various reactions to it, thus generating more colorful images than one can shake a kaleidoscope at. Such studied eclecticism makes postmodernism inconsistent from a rational point of view and renders a strictly rational discussion of it practically impossible.

These inconsistencies are nicely reflected in the widespread sentiments toward chaos theory, which is frequently hailed as the latest 'paradigm shift" and advertized as "postmodern mathematics."

But unlike what some postmodern "mathematicians" seem to believe, even chaos can be conceived only against the background of the regulative idea of determinism. Simply said, chaos theory studies dynamical systems with deterministic time-evolution. The history of a "chaotic" system *is* determined, in theory; it is just that *our* ability to predict mathematically the behavior of the system is sharply limited in a certain technical sense.

More problematically, the mathematics behind chaos theory employs metaphysical idealizations such as points, systems, sets, structures, infinity, infinite limit processes that are overtly assumed to be capable of completion, and system states that are described as Platonist points in a Platonist infinite-dimensional universe. It deals with abstract functions, the totalizing logic of identity, binary thinking, ultimate grounds of justification, and many other things the mere mention of which should induce any postmodern acolyte to make the sign of the cross.

If anything, it seems that chaos theory, particularly as a model of social dynamics or creative processes, should be *questioned* by postmodernism. At best, such applications carry along the heavy "metaphysical" burden of pre-postmodern mathematics. At worst, they implicitly presuppose a rather crude form of determinism.

But perhaps postmodernism is not allergic to social physics after all. Perhaps "postmodern mathematics" is simply an ideological vehicle used by various academics to bring about Foucault's wish for the return to "happy positivism." (In the age of chaos, a small action like switching a channel can cause the fall of the Italian government. At long last, scientists have explained this strange phenomenon.)

Thus, in despite of some similarities with the culture of romanticism, the most popular versions of what passes for postmodern thought seem to be little more than bold variations on the famous "logocentric" tenet: To be is to be the value of a variable. The variable is now chaotic and random, but the principle is the same. Languages speak, structures mean, and changes occur courtesy of a mysterious "power-in-general" that belongs to no one in particular, which is to say that we are dealing with a kind of functionalism. One could perhaps call it dysfunctionalism. Dysfunctionalism *chic*.

Postmodern culture (whatever one understands by this vague term) does not seem to mark a decisive break from anything that came before it. On the contrary, it is, among many other things, a case of formalism taken to unfathomable extremes, enriched by a distorted version of ideas of some people who criticized formalism a long time ago. In its almost complete lack of mathematical-historical awareness, it managed to combine formalist reductionism and intuitionist radicalism. In short, it combines precisely the worst features of both, for the most part blissfully unaware of either.

On this issue, it is difficult to disagree with Guattari's comment in his article "The postmodern impasse":

> [P]ostmodernism is nothing but the last gasp of modernism; nothing, that is, but a reaction to and, in a certain way, a mirror of the formalist abuses and reductions of modernism from which, in the end, it is no different. [. . .] However, [this] is nothing to rejoice about, as postmodernists seem to think. The question, rather, ought to be: how can we escape this dead end?[24]

It seems possible, then, that postmodernism is merely a carnival gone wrong. Merry revelers are experiencing difficulties unzipping the costumes they used in order to parody the objects of their objections. What remains is a bungling procession of distorted mirror images of science, mathematics, romantic idealism, positivism, materialism, Marxism, humanism, structuralism, cultural studies, and finally of Derrida's philosophy of deconstruction.

But this is nothing to rejoice about. The question is: How can one escape this dead end? I would like to believe that mathematical analogies outlined in this book—regardless of how speculative and informed by my limitations they may be—open up the possibility of a dialog that might proceed under a designation less evocative of violence than "science wars." Mathematics, that much seems to me beyond reasonable doubt, is entangled with continental philosophy in a complicated way. I hope to have indicated that it is not entirely irrelevant even to the diverse collection of theories somewhat arbitrarily lumped together under the heading "postmodern thought."

It would certainly be nonsensical to claim that mathematicians have "done it all," that most of twentieth-century thought secretly revolves around the debate between intuitionism and formalism, and that the fluttering wings of mathematicians' imaginations cause storms in distant regions of philosophy. Nevertheless, the broad cultural significance of mathematics (which I emphasized, or overemphasized) permits us to view apparently disconnected endeavors in a light perhaps a little softer than the blinding flashes of "science wars." The belief in the possibility and utility of a hermeneutical dialog may be an antiquated, romantic, humanist illusion, but it is one that I hold dear.

NOTES

Chapter 1

1. For a sample of amusing (but not terribly illuminating) assertions about the latest "paradigm shift"—the effect of chaos theory, fractal geometry, and something called the new nonlinear or fuzzy math—see William V. Dunning, *The Roots of Postmodernism* (Englewood Cliffs, N.J., Prentice-Hall, 1995), especially the chapter "The Dragons of Postmodern Art and Science."

2. Bruno Latour, *We Have Never Been Modern*, trans. Catherine Porter (Cambridge, Mass.; Harvard University Press, 1993).

3. "The mathematical, in the original sense of learning what one already knows, is the fundamental presupposition of all 'academic' work. [. . .] Therefore, we must now show in what sense the foundation of modern thought and knowledge is essentially mathematical" (Martin Heidegger, *Basic Writings*, ed. by David Farrell Krell ([New York, Harper & Row, 1977], p. 254).

4. Byron, *Manfred* (1817), quoted from "Byron", *Encyclopedia Britannica Online*, <http://10.9.4.102/bol/topic?eu=18687&sctn=1>

Chapter 2

1. Quoted from *Lakatos Philosophical Papers*, Vol. 2, *Mathematics, Science and Epistemology* (Cambridge, Cambridge University Press, 1978), pp. 78–9. My brief discussion of Descartes follows Lakatos's much more extensive analysis of the Cartesian Circuit.

2. See "Newton, the Man," in *The Collected Writings of John Maynard Keynes*, Vol. 10 (London, Macmillan Press, 1972), pp. 363–4, 374.

3. Immanuel Kant, *Critique of Pure Reason*, trans. by Norman Kemp Smith (New York, St. Martin's Press, 1978), A128–30.

4. Ibid., A602.

5. Ibid., A55, B80.

6. Ibid., B153.

7. Ibid., A475–6, B503–4 (italics added).

8. Isaiah Berlin, *The Roots of Romanticism* (Princeton, N.J., Princeton University Press, 1999), pp. 94–5.

9. Novalis, *Schriften* (Stuttgart, Kohlhammer, 1960), vol. 2, p. 112.

10. Fichte, *Sämtliche Werke*, Vol. 2 (Berlin, Veit, 1845–6), p. 263. Quoted from Berlin (1999), p. 89.

11. J.W. Goethe, *Faust I*, in: *Faust und Urfaust* (Wiesbaden, Dietrich'sche Verlagbuchhandlung, 1948), p. 48.

12. *Hegel's Philosophy of Right*, trans. by T.M. Knox (Oxford, Oxford University Press, 1967), preface.

13. G.W.F. Hegel, *Phenomenology of Mind*, trans. J.B. Baillie (London, 1931), Preface.

Chapter 3

1. Kant, *Critique of Pure Reason*, B16–7.

2. Gauss's letter to Bolyai, Sr. Quoted from Michael Detlefsen, "Philosophy of Mathematics in the Twentieth Century," in: Stuart Shanker (ed.), *Philosophy of Science, Logic and Mathematics in the 20th Century*, Routledge History of Philosophy, Vol. 9 (New York, Routledge, 1996), pp. 53–4.

3. See Albert A. Blank, "The Luneburg Theory of Binocular Space Perception," in: Sigmund Koch (ed.), *Psychology: A Study of a Science*, Vol. 1 (New York, McGraw-Hill, 1959), pp. 395–496, esp. p. 414.

4. Morris Kline, *Mathematics: The Loss of Certainty* (New York, Oxford University Press, 1980), p. 76.

5. Kant, *Critique of Pure Reason*, A78, B103.

6. Ibid., A831. See also Albert Einstein, "Physics and Reality," *Journal of the Franklin Institute*, **221** (1936), 313–47, where he states that physicists "cannot proceed without considering critically a more difficult problem, the problem of analyzing the nature of everyday thinking."

7. Kant, *Critique of Pure Reason*, Avii–viii (italics added).

8. See Erwin Panofsky, *Galileo as a Critic of the Arts* (The Hague, Nijhoff, 1954), p. 13.

9. For an interesting though perhaps more controversial comparative analysis of Kant and Bentham, see Slavoj Žižek, *Tarrying with the Negative: Kant, Hegel, and the Critique of Ideology* (Durham, N.C., Duke University Press, 1993), pp. 86–9. Citations of Bentham given in the text, as well as the comparison of his views to logical positivism, come from Žižek's book. For more on Bentham's theory of fictions, see Ross Harrison, *Bentham* (London, Routledge & Kegan Paul, 1983).

10. An interesting literary example of shattered geometrical illusions appears in the novel *The Hearing Trumpet* (1976) by the English-born American writer and painter Leonora Carrington. The story is about an old lady who enters a nursing home and finds that there are only a few pieces of actual furniture in the entire place. The rest of the furniture is painted, that is, not three dimensional. This has a "strangely depressing effect." Once her 3D illusions are broken, which

is represented in the novel by an earthquake that opens a crack in the building, she descends through this fissure into a very bizarre underworld. She ends up as a split personality who engages in an act of self-referential cannibalism, watching herself making a stew out of herself. If a surreal novel like *The Hearing Trumpet* could be said to have a "moral," it would perhaps be this: Some formative fantasies are necessary. They may be illusory from the scientific point of view, but without them the world (as I know it) might collapse into a nightmarish rubble.

11. G.W. Leibniz, *Quid sit idea*, in: Carl Gerhard (ed.), *Die Philosophischen Schriften von G.W. Leibniz*, Vol. 7 (Berlin, Weidmann, 1875), pp. 263–4. English translation quoted from Umberto Eco, *The Search for the Perfect Language* (Oxford, Blackwell, 1995), p. 284.

12. G.W. Leibniz, *Historia et commendatio*, in Gerhard (1875), Vol. 7, p. 184. English translation quoted from Eco (1995), p. 276. Eco comments that Leibniz's analyses "led him to conclude that an alphabet of primitive thought could never be formulated," except as an "elegant artifice." For further details, see Eco (1995), Chap. 14.

13. "Our reason is formed only through fictions. We incessantly seek and create for ourselves a One in Manifold and give it form; out of this arise concepts, ideas, ideals" (J.G. Herder, *Sämtliche Werke*, Vol. 17 [Berlin, Weidmann, 1877], p. 485). English translation quoted from Manfred Frank, *What Is Neostructuralism?* (Minneapolis, Minnesota University Press, 1989), p. 129.

14. Johann Georg Hamann, *Sämtliche Werke*, Vol. 3 (Vienna, 1949–57), p. 225. English translation quoted from Berlin (1999), p. 44.

15. Immanuel Kant, *Conflict of Faculties* (New York, Abaris Books, 1979), pp. 55–7.

16. Carl Boyer, *A History of Mathematics*, 2nd ed., rev. by Uta Merzbach (New York, Wiley, 1989), p. 520.

17. The term "intellectual intuition" was used by several romantic philosophers in different ways, but it apparently meant something other than psychological introspection. Fichte actually complained about being misconstrued in such a manner: "The more a determinate individual can think himself *away*, the closer does his empirical self-consciousness approximate pure self-consciousness" (*Sämtliche Werke*, Vol. 1, p. 244, italics added). Friedrich Schelling describes intellectual intuition similarly, as an awareness of the activity "of self-consciousness *in general*" (*Sämtliche Schriften* Vol. 2, [Stuttgart, Cotta, 1856–61], p. 374, italics added).

18. Jean Dieudonné, *Mathematics—The Music of Reason* (New York, Springer, 1992), p. 203.

Chapter 4

1. Schelling, *Sämtliche Schriften*, 1/10, pp. 11–2. English translation quoted from Frank (1989), p. 299.

2. Berlin (1999), p. 105.

3. Quoted from Berlin (1999), p. 104 (German original untraced; possibly a paraphrase).

4. Friedrich Nietzsche, *The Will to Power* (New York, Random House, 1967), p. 449.

5. "Art is the Tree of Life ..." comes from Blake, "Lacoon," aphorisms 17, 19. Quoted from Berlin (1999), p. 50. "God forbid ..." is from William Blake, *Notes on Reynold's Discourses*, as cited on the Mathematical Quotation Server, http://math.furman.edu/~mwoodard/mqs/mquot.shtml.

6. Quoted from "Schlegel, Friedrich von", *The Cambridge Dictionary of Philosophy* (New York, Cambridge University Press, 1995), p. 716.

7. "Nobody conceives in a given word exactly what his neighbor does, and the ever so slight variation skitters through the entire language like concentric ripples over the water. All understanding is simultaneously a non-comprehension, all agreement in ideas is at the same time a divergence" (Wilhelm von Humboldt, *Linguistic Variability and Intellectual Development* [Philadelphia, University of Pennsylvania Press, 1972], p. 43).

8. Novalis, *Heinrich von Ofterdingen*, Part 2, Vol. 3, p. 434. English translation quoted from Berlin (1999), p. 104.

9. Brouwer, "The Structure of the Continuum," lecture given in Vienna, March 14, 1928. Quoted from Mancosu (1998), p. 54.

10. These are free variations on Bergson's theme. For more details on Bergson's outlook, see Henri Bergson, *Time and Free Will* (New York, Humanities, 1971) and Lesek Kolakowski, *Bergson* (Oxford, Oxford University Press, 1985).

11. Quoted from Kolakowski, *Bergson*, p. 129.

12. Kant, *Critique of Pure Reason*, A526.

13. Schelling, *Sämtliche Schriften*, 2/2, p. 206. English translation quoted from Frank (1989), p. 423. The idea of relaxing the rigidity of causal chains—as in Brouwer's choice-sequences—to make room for acts of mental construction can also be found in the work of American philosopher Charles Sanders Peirce, ca. 1903: "[B]y supposing the rigid exactitude of causation to yield [...] we gain room to insert mind into our scheme, and to put it into the position which, as the sole self-intelligible thing, it is entitled to occupy, that of the *fountain of existence*" (*Collected Papers*, ed. by C. Hartstone and P. Weiss, Vol. 6 [Cambridge, Mass., Harvard University Press, 1953], pp. 42–3, italics added).

14. L.E.J. Brouwer, "Intuitionistische verzamelingsteer," *KNAW Verslagen*, **29** (1921), 797–802. English translation quoted from Paolo Mancosu, *From Brouwer to Hilbert* (New York, Oxford University Press, 1998), p. 23.

15. Ludwig Wittgenstein, *Philosophical Investigations* (Oxford, Basil Blackwell, 1953), para. 456.

16. Heidegger, *Basic Writings*, p. 254.

17. Martin Heidegger, *Being and Time* (New York, Harper & Row, 1962), p. 121.

18. Heidegger, "Letter on Humanism," in *Basic Writings*, p. 227.

19. Both Brouwer and Heidegger criticize man's domination by means of reason, that is, science as "knowledge for mastery," and emphasize the limits of formal-logical reasoning. But Heidegger objects to the alleged sovereignty of the idealist subject in "creating" truth. For him—and "postmodernism" for the most

part simply takes over Heidegger's view—this notion represents the historical culmination of man's drive for mastery. He will therefore try to "overturn" idealism by showing that subjectivity is itself *derived* from the structure of existential understanding ("the space in which always already move"). In particular, intuition itself is "derived." The argument is controversial, but part of it is interesting and relevant to mathematics. Heidegger argues that intuitions are influenced by formalizations and conversely, but that neither of the two is "the origin" of mathematical concepts. Thus, the controversy between intuitionists and logicians is not likely to be resolved in a simple manner: "One of [the] burning questions concerns the limits and justification of mathematical formalism in contrast to the demand for an immediate return to intuitively given nature. [. . .] The question cannot be decided by way of an either/or, either formalism or immediate return to intuitive determination of things; for the nature and direction of the mathematical project participate in deciding their possible relation to the intuitively experienced and vice versa." (*Basic Writings*, p. 270).

20. Indeed, Heidegger's awareness of the debate about the nature of mathematical knowledge is made clear already in the first chapter of *Being and Time*: "The discipline which is seemingly the strictest and most securely structured, mathematics, has experienced a 'crisis in its foundations'. The controversy between formalism and intuitionism centers on obtaining and securing primary access to what should be the proper object of this science" (*Basic Writings*, p. 51).

21. Brouwer, *Life, Art, and Mysticism* (1905). See Brouwer, *Collected Works*, ed. by A. Heyting (Amsterdam, North-Holland, 1975), p. 5. Brouwer's quotes in the following paragraph are from the same source.

22. Brouwer, *On the Foundations of Mathematics* (1907). See *Collected Works*, p. 130.

23. Brouwer, *Collected Works*, p. 132.

24. *Brouwer's Cambridge Lectures on Intuitionism*, ed. by D. van Dalen (Cambridge, Cambridge University Press, 1981), pp. 4–5.

25. Brouwer, "Mathematik, Wissenschaft und Sprache" ("Mathematics, Science, and Language"). Lecture given in Vienna on March 10, 1928. English translation quoted from Mancosu (1998), p. 48.

26. Nietzsche, *The Gay Science* (New York, Random House, 1974), aphorism 121.

27. Brouwer, "Mathematics, Science and Language," in: Mancosu (1998), p. 45.

28. Ibid., p. 47.

29. Ibid., p. 48.

30. Brouwer, *Life, Art, and Mysticism*, p. 2. Although the notion of a deeper "spirit" inaccessible to language and reason may be called a "romantic" one, it would perhaps be more accurate to say that it gained philosophical prominence with the rise of romanticism. For instance, one finds related ideas in Luther's indictment of the "harlot reason," in the Sufi poet Jelaludin Rumi's statement that language cannot reach the one who lives "within you," and, last but not least, in the inward-looking doctrine of gnostic mysticism.

31. Brouwer, "Mathematics, Science, and Language," in: Mancosu (1998), p. 48.

Chapter 5

1. See Mancosu (1998), p. 135.
2. Emile Borel, "À propos de la recente discussion entre M. R. Wavre et M. P. Levy," *Revue de Métaphysique et de Morale*, **34** (1927), 271–6. English translation quoted from Mancosu (1998), p. 296.
3. See D. van Dalen, "Four Letters from Edmund Husserl to Hermann Weyl," *Husserl Studies*, **1** (1984), 1–12.
4. See Mancosu (1998), pp. 94–5.
5. Edmund Husserl, *Cartesian Meditations: An Introduction to Phenomenology* (The Hague, Nijhoff, 1960), pp. 22–3.
6. "It is a flow of phenomena [. . .] a time that no chronometer can measure" (Edmund Husserl, *Phenomenology and the Crisis of Philosophy* [New York, Harper & Row, 1965], pp. 107–8).
7. Jacques Derrida, *Edmund Husserl's Origins of Geometry* (Stony Brook, Nicholas Hays, 1978), pp. 137, 153. Elsewhere, Derrida simply says that pure difference comes to divide self-presence.
8. The term *logistique* is used in Plato's sense here and refers to the practical skill of calculation as used by businessmen and men of war, which is of little interest to philosophers. Plato regarded it as inferior to "true" mathematics, that is, geometry and arithmetic. (See *The Republic of Plato*, trans. by Francis Macdonald Conford [New York, Oxford University Press, 1945], p. 241.) Poincaré was only too happy to explain this in detail: "The same word is used in the *École de Guerre* to designate the art of the quartermaster, the art of moving and quartering troops" (*Science and Method* [New York, Dover, 1952], p. 156).
9. The sequence of colorful comparisons, employed throughout Poincaré's writing on the topic, culminates with the following description of what would now be called an automated theorem-proving system: "[I]f we prefer, we might imagine a machine where we should put axioms at one end and take theorems at the other, like that legendary machine in Chicago where pigs go in alive and come out transformed into hams and sausages" (*Science and Method*, p. 147).
10. See Antonio R. Damasio, *Descartes' Error: Emotion, Reason, and the Human Brain* (New York, Avon Books, 1994), pp. 188–9.
11. "On the Foundations of Geometry," *Monist*, **9** (1898), 1–43. Quoted from Arthur Miller, *Imagery in Scientific Thought* (Cambridge, Mass., MIT Press, 1986), p. 20 (italics added).
12. "Les géometries non-euclidiènnes," *Rev. Générale Sci. Pures Appl.*, **2** (1891), 769–74. Quoted from Miller (1986), p. 19.
13. Ibid.
14. Whether he was aware of it or not, Poincaré seems to have reinvented in relatively clear mathematical terms an argument that had been put forward by Fichte and Schelling, namely, that identity cannot be formally deduced and so must always be presupposed. For example, in his *Basis of the Entire Theory of*

Science (1794), Fichte says that "[i]f the proposition $A = A$ is certain, then the proposition *I am* must also be certain" (*Sämtliche Werke*, Vol. 1, p. 95). Since Fichte certainly knew that the proposition "I am" is not *logically* certain, it follows that attribution of identity always involves a hypothetical judgment. The idea that identity is always presupposed was taken up by Schelling, who argues that absolute self-identity is not formally deducible and takes it as the irreducible first principle of philosophy.

15. For Heidegger, formulation of mathematical axioms represents only one aspect of the mathematical project. Formalization brings to surface the forms of knowledge already *anticipated* through the "clearing of Being" (an "existential understanding" or interpretation of the world, which comes prior to both formalization and intuition): "[T]he mathematical project is the anticipation of the essence of things, of bodies; thus the basic blueprint of the structure of every thing and its relation to every other thing is sketched out in advance" (*Basic Writings*, p. 268). These formalizations are neither complete nor immutable; they change as *understanding* changes through history.

16. See Charles Misner, Kip Thorne, and John Wheeler, *Gravitation* (San Francisco, Freeman, 1970), p. 1205. As to the dimensionality of space, Poincaré's view is perhaps a little surprising: Space could have as many dimensions as there are muscles in our bodies. See Miller (1986), p. 22.

17. For Nietzsche's quote on logic, see *The Complete Works of Friedrich Nietzsche*, ed. by Oscar Levy, Vol. 15 (New York, Macmillan, 1909–13), p. 33. For Heidegger's statement, see *Basic Writings*, p. 268 (italics added).

18. "Sur la nature du raisonnement mathematique," *Revue de Métaphysique et de Morale*, **2** (1894), 371–84. English translation quoted from Miller (1986), p. 60. This is a matter of controversy among experts. Swiss physician and psychologist Jean Piaget, who disagreed with the American linguist Noam Chomsky on the innateness of language, also disagreed with Poincaré over innateness of the group of continuous motions. See Jean Piaget, *The Psychology of the Child* (New York, Basic Books, 1969). However, several recent studies question Piaget's doctrine, and a number of psychologists have agreed with Poincaré at least partially. For researchers outside of Piaget's school, the question is not whether continuity is "constructed" by the mind, but rather how this construction proceeds. This is a serious problem, of interest not only to psychology but also to philosophy. For a study of the "construction" of the continuity of motion, see Paul A. Kolers, *Aspects of Motion Perception* (Oxford, Pergamon Press, 1972).

19. From Poincaré's 1912 lecture at the University of London; see Miller (1986), p. 24.

20. For more about this unusual world of codes and clones, see the work of the California-based French philosopher Jean Baudrillard, especially *Simulation* (New York, Columbia University Press, 1981).

21. Poincaré explored the difficulties with justifying mathematical induction in his critiques of reductions of mathematics to logic or other purely formal systems. For Poincaré, it is not possible to think without already using a form of mathematical induction—making an inductive "leap" from n to $n + 1$—even if it is in the form of employing a grammatical plural: "The fact is that it is impos-

sible to give a definition without enunciating a phrase, and difficult to enunciate a phrase without putting in a name of a number, or at least the word 'several,' or at least a word in the plural. Then the slope is slippery, and at each moment one risks falling into the *petitio principii*" (*Science and Method*, p. 155). Husserl's disciple Oskar Becker, who seems to have been influenced by Weyl and Heidegger, refines Poincaré's objections in his 1927 article "Mathematische Existenz." See Mancosu (1998), pp. 165-7. The issue has been addressed even by the "postmodern" psychoanalyst Jacques Lacan, who relates it to the question of subjectivity. Lacan notes that the formula "$n + 1$" contains no less than "the question of the subject." If Lacan's mathematical metaphors were less obscure, one could perhaps search for a connection with Poincaré in Lacan's piece on mathematical induction. Unfortunately, his writing on the topic is difficult to decipher. For a quick look at this uncharted nebula, and amusing commentary, see Alan Sokal and Jean Bricmont, *Fashionable Nonsense: Postmodern Intellectuals' Abuse of Science* (New York, Picador, 1998), pp. 28-31.

22. Warren Goldfarb, "Poincaré against the Logicists," in: P. Kitcher and W. Aspray (eds.), *History and Philosophy of Modern Mathematics* (Minneapolis, Minnesota University Press, 1988), p. 78 (italics added).

23. Ferdinand de Saussure, *Cours de linguistique générale*, Edition Critique, ed. by Rudolf Engler, Vol. 2 (Wiesbaden, Harrasowitz, 1967-74), p. 23. English translation quoted from Frank (1989), p. 426. Engler's critical edition is henceforth quoted as *Edition Critique*.

Chapter 6

1. "Die Grundlegung der elementaren Zahlentheorie," *Mathematische Annalen*, **104** (1931), 485-94. English translation quoted from Mancosu (1998), p. 170. Hilbert mentions the relevance of Kant's position already in his lectures from 1922-3, but the lecture notes were not published until much later.

2. "Neubegründung der Mathematik. Erste Mitteilung," *Abhaundlungen aus dem Mathematischen Seminar der Hamburgischen Universität*, **1** (1922), 157-77. English translation quoted from Mancosu (1998), p. 198.

3. Quoted from Berlin (1999), p. 89.

4. From Kant, *Critique of Pure Reason*, as quoted in Detlefsen (1996), p. 79. See also pp. 76-81 for a more detailed discussion of Hilbert versus Kant.

5. Herman Weyl, "The Current Epistemological Situation in Mathematics," in: Mancosu (1998), p. 140.

6. Edmund Husserl, *The Crisis of European Sciences and Transcendental Phenomenology* (Evanston, Northwestern University Press, 1970), subsection II.9.g. The subsection has the heading "Emptying of the meaning of mathematical natural science through technization." In the same section, Husserl mentions "intuitionistic deepening" of the project of mathematics, and discreetly laments the prejudices of "positive sciences" that prevented intuitionism from finding wider acceptance.

7. Such a dispersion of language games into incommensurable localities is supposed to be one of the guiding visions of postmodern thought, although the

notion is typically transferred into the domain of social studies. For example: "Social subject itself seems to dissolve in this dissemination of language games" (Jean-Francois Lyotard, *The Postmodern Condition: A Report on Knowledge* [Minneapolis, University of Minnesota Press, 1984], p. 66).

8. For more on the exchange between Hilbert's school and Nelson's neo-Kantian group, see Mancosu (1998), pp. 173–5.

9. "Diskussionbemerkungen zu dem zweiten Hilbertschen Vortag über die Grundlagen der Mathematik," *Abhaundlungen aus dem Mathematischen Seminar der Hamburgischen Universität*, **6** (1928), 86–8. English translation quoted from J. van Heijenoort, *From Frege to Gödel* (Cambridge, Mass., Harvard University Press, 1967), p. 484. See also Mancosu (1998), p. 81.

10. From Gödel's correspondence, as quoted in Solomon Feferman, *In the Light of Logic* (New York, Oxford University Press, 1998), p. 159.

11. Louis Couturat, *Opuscules et fragments inédits de Leibniz* (Paris, Alcan, 1903), p. 28. Quoted from Eco (1995), p. 277.

12. From Gödel's correspondence, as quoted in Feferman (1998), p. 158.

13. For more on Gödel's work and philosophical outlook, see Feferman (1998), Chaps. 6–8.

Chapter 7

1. Sociologically speaking, it seems that embracing Cavaillès's formalism functioned as a rebellious gesture among young intellectuals who were bored with Bergson's ineffable flux and Sartre's existentialist nausea and anxiety. Later, I examine more closely the shape that this antiromanticist rebellion took. It is commonly referred to as "structuralism."

2. Jean Cavaillès, *Sur la logique et la théorie de la science* (Paris, Vrin, 1987), p. 23. English translation quoted from John Lechte, *Fifty Contemporary Thinkers* (London, Routledge, 1994), p. 18. Cavaillès's notion that the truth of a theorem resides *in* the demonstration, considered as a movement within the structure of science itself, seems to be related to Hegel. At least, Hegel says something vaguely similar: "The true form in which truth exists can only be the scientific system of the same" (*The Phenomenology of Mind*, trans. by J.B. Baillie [London, 1931], p. 70).

3. Michel Foucault, *The Archeology of Knowledge* (New York, Pantheon, 1971), p. 182.

4. See footnote 48a in Kurt Gödel, "Über formal uneinscheindbare Sätze der Principia Mathematica und verwandter Systeme I," *Monatshefte für Mathematik und Physik*, **38** (1931), 173–98. English translation quoted from Feferman (1998), p. 161.

5. Foucault, *The Archeology of Knowledge*, p. 224.

6. Ibid., p. 182. Traces of Heidegger's influence can be detected in this passage. Foucault prefers the term "discursive practice" to Heidegger's more traditional term "method," but their points are quite similar: "Method is not one piece of equipment of science among others but *the primary component out of which is first determined what can become an object and how it becomes an object*" (Heidegger,

"Modern Science, Metaphysics, and Mathematics," in *Basic Writings*, p. 277, italics added).

7. "It is no longer possible to think in our day other than in the void left by man's disappearance. For this void does not create a deficiency; it does not constitute a lacuna that must be filled. It is nothing more, and nothing less, than the unfolding of a space in which it is once more possible to think" (Foucault, *The Order of Things: An Archaeology of the Human Sciences* [New York, Vintage Books, 1970], p. 342).

8. According to Foucault, human sciences can become sciences only when they apply the methods of formal linguistics and study the language as something that is supposedly *exterior* to human beings: "In linguistics, one would have a science perfectly founded in the order of positivities exterior to man (since it is a question of pure language)" (*The Order of Things*, p. 381).

9. "With Brouwer, mathematics gains its highest intuitive clarity [. . .]. But, full of pain, the mathematician sees the greatest part of his towering theories dissolve into fog" (quoted from Mancosu [1998], p. 80).

10. Foucault, *The Order of Things*, p. 364 (italics added).

11. Ibid., p. xiii.

12. More precisely: "Since man was constituted at a time when language was doomed to dispersion, will he not be dispersed when language regains its unity?" (Foucault, *The Order of Things*, p. 385). It is quite remarkable that Foucault, except for the obligatory doom-writing tenor with regard to the subject of humanity, goes against the received wisdom of popular postmodernism, which sees fragmentation of discourse as a natural, inevitable, and presumably eternal state of affairs. Lyotard, for instance, claims that in "the society of the future" there "are many different language games—a heterogeneity of elements. They only give rise to institutions in patches—local determinism" (*The Postmodern Condition*, pp. xxiii–xxiv). (Interestingly enough, Lyotard's celebration of "local determinism" was commissioned by the government of Quebec.)

13. Foucault, *The Order of Things*, p. 382.

14. It is difficult to be certain about this, despite the ubiquity of "cyborg manifestos" and articles about a new species that calls itself "telematic." But Lacan, for instance, argues that there would still be a *subject*—an "I" (*Je*), which is Lacan's term for the id—even if all living beings disappeared from the earth, leaving only a camera to "record the image of the Café Flore in the process of crumbling in complete solitude." There would even be *consciousness*, because "consciousness is produced whenever there is [. . .] a surface that is such that it can produce what is called an image." Now, if a camcorder can be considered as being conscious or having an "I," why not a cleverly programmed computer? After all, as Lacan says, "[T]he most complicated machines are made only with words" (Lacan, *Séminaire, Livre II, Le moi dans la théorie de Freud et dans la technique de la psychanalyse* [Paris, Seuil, 1978], pp. 62–5. English translation quoted from Frank [1989], pp. 311–3). I return to this problem later in a less laconic mood.

15. "Jean Paul Sartre répond," *L'Arc*, **30** (1966), 87. English translation quoted from Frank (1989), p. 103. A similar depiction of Foucault's archeology

came from the pen of Jean Piaget: "The message of this 'archeology' is, in short, that reason's self-transformations have no reason and that its structures appear and disappear by fortuitous mutations and as a result of momentary upsurges" (*Structuralism* [New York, Harper & Row, 1970], p. 134).

16. "[T]he real truth—the glaring, sober truth that resides in delirium" (Gilles Deleuze and Félix Guattari, *Anti-Oedipus* [New York, Viking Press, 1977], p. 4). The idea seems to be that ossified systems can be transgressed only by means of chaotic, undirected, "horizontal" thinking. For as soon as my thoughts have *an* order, I am not innovative: I am already "coded," caught in *a* system's perfidious web. Thus, reason becomes torture for Foucault, while social consensus is suspected by Lyotard to be a form of terrorism. So "we" must fight order not in the name of a different and better order—but fight it *as such*. Curiously, and perhaps a little embarrassingly for these revolutionaries, a similar Tao of Chaos has been embraced by the management guru Tom Peters. In his *Thriving on Chaos: Handbook for a Management Revolution* (Toronto, Random House of Canada, 1987), Peters "argues" for a liberation from reason in the name of senseless and endless production of change. Supporting his fulminations against Newton and Descartes by vacuous references to Jane Smiley and quantum physics, Peters recommends the management principles of Genghis Khan's Mongol hordes, which he sees both as "groups of freelance bandits" and as basic organizational units of the new "horizontal" corporation. Peters is by no means alone in this crusade: The concept of "scientific" management has been under attack for a few decades. For an illuminating tour of pseudorebellious antirationalism in business culture and management literature, see Bill Boisvert, "Apostles of the New Entrepreneur: Business Books and the Management Crisis," in: Thomas Frank and Matt Weiland (eds.), *Commodify Your Dissent* (New York, Norton, 1997), pp. 81–98. It appears that certain postmodern proposals are not at all incompatible with such forms of authoritarian thinking. In light of this compatibility, Lyotard's assertion that "postmodern knowledge is not simply the tool of the authorities" (*The Postmodern Condition*, p. xxv) seems a bit premature.

17. Deleuze and Guattari, (1977), p. 76.

Chapter 8

1. Heidegger traces a version of this idea back to Newton. According to Heidegger, with Newton's doctrine of motion the "concept of place itself is changed: place is no longer where the body belongs according to *its inner nature*, but only a position *in relation to other* positions" (*Basic Writings*, p. 263, italics added).

2. Ferdinand de Saussure, *Course in General Linguistics* (New York, McGraw-Hill, 1966), pp. 111–2.

3. Ibid., p. 112.

4. Ibid., p. 120. See also *Cahiers Ferdinand de Saussure*, **15** (1957), p. 93: "In *langue*, there is nothing but differences, no positive quantity."

5. "[A]ll thought, therefore, must necessarily be in signs" (Peirce, *Collected Papers*, Vol. 5, para. 251). Schleiermacher wrote that "nobody can think with-

out words. Without words thought is not yet completed and clear" (*Hermeneutik und Kritik*, ed. by Manfred Frank [Frankfurt, Suhrkamp, 1977], p. 77). English translation quoted from Frank (1989), p. 212.

6. Saussure, *Course in General Linguistics*, p. 122.

7. For Weyl's quote, see Hermann Weyl, *Philosophy of Mathematics and Natural Science* (New York, Atheneum Press, 1963), pp. 25–6. Source for Poincaré's quote: J.R. Newman (ed.), *The World of Mathematics* (New York, Simon & Schuster, 1956), as cited on the Mathematical Quotation Server, http://math.furman.edu/~mwoodard/mqs/mquot.shtml.

8. Richard Dedekind, *Was sind und was sollen die Zahlen?* (Braunschweig, Vieweg, 1888), para. 73. English translation adapted from Detlefsen (1996), p. 104.

9. The analogy between mathematics and structural linguistics does not seem to be entirely the product of my mathematical bias. For instance, Danish structural linguist Louis Hjemslev explicitly acknowledges the formative influence of mathematics on his theory, which he calls *glossematics*. Glossematics, according to Hjemslev, is a science that takes the "immanent algebra of language" as its object of study; it is "an algebra of language operating with unnamed entities" (*Prolegomena to a Theory of Language* [Madison, University of Wisconsin Press, 1963], pp. 79–80).

10. See Frank (1989), p. 8.

11. See Engler's *Edition Critique*, Vol. 1, p. 243, and *Cahiers Ferdinand de Saussure*, **15** (1957), 89. Quoted from Frank (1989), pp. 441–2.

12. The collective imposes some inertia on language, but does not determine its meaning: "The [law of language] is general but not imperative. Doubtless it is imposed on individuals by the weight of collective usage [. . .]." And yet, "in language no force guarantees the maintenance of regularity when established on some point" (Saussure, *Course in General Linguistics*, p. 92).

13. Claude Lévi-Strauss, *Structural Anthropology* (New York, Doubleday, 1967), p. 227.

14. See chap. 6, n. 10.

15. Claude Lévi-Strauss, *The Raw and the Cooked* (New York, Harper & Row, 1970), p. 12.

16. Lévi-Strauss, *Structural Anthropology*, p. 226.

17. For Bourbaki, mathematics is based on three fundamental types of structures: order, algebra, and topology. This particular categorization is in a sense analogous to Piaget's theory of the development of mathematical concepts in children, at which he arrived independently from Bourbaki. See Reuben Hersh, *What Is Mathematics, Really?* (New York, Oxford University Press 1997), pp. 225–7.

18. For Rota's (apparently controversial) article, see Gian-Carlo Rota, *Indiscrete Thoughts* (Boston, Birkhäuser, 1997), pp. 89–103. Thom's polemical paper on "modern mathematics" is reprinted in Thomas Tymoczko (ed.), *New Directions in the Philosophy of Mathematics*, 2nd ed., rev. (Princeton, N.J., Princeton University Press, 1998), pp. 67–78.

19. Roland Barthes, *Critical Essays* (Evanston, Ill., Northwestern University Press, 1979), p. 213.

20. See Vladimir Propp, *Morphology of the Folktale* (Austin, University of Texas Press, 1968), and Algridas Julien Greimas, *Structural Semantics* (Lincoln, University of Nebraska Press, 1966).

21. Benjamin Lee Whorf, "Science and Linguistics," *Technology Review*, **42** (1940), reprinted in John B. Carroll (ed.), *Language, Thought, and Reality: Selected Writings of Benjamin Lee Whorf* (Cambridge, Mass., MIT Press, 1956), pp. 207–19 (italics original).

22. There is a great deal of "dialectical" ambiguity in these claims, and they have been interpreted in different ways. For example, in preface to *A Contribution to the Critique of Political Economy*, Marx says that the structure of productive relations "conditions" (*bedingt*)—rather than determines—mental life.

23. Thus the famous notion of the "prison-house of language." But Nietzsche seems cautious and attributes the "wonderful resemblance of Indian, Greek, and German philosophizing" to "the unconscious *domination and guidance* of similar grammatical functions" (*The Complete Works of Friedrich Nietzsche*, New York, Macmillian, 1909–13, Vol. 12, p. 29, italics added). The choice of words is interesting. For instance, laws of physics certainly "guide" and "dominate" my movements, but they do not determine my postal code. In fact, Nietzsche continues to say (loc. cit.) that these dominant grammatical functions serve to *remove* "certain other possibilities of world interpretation." That is quite different from linguistic determinism. On the other hand, Nietzsche also describes language as a mobile army of metaphors that "after a long usage *seem to a nation fixed, canonic and binding*" (*The Complete Works*, Vol. 2, p. 180, italics added). (See also chap. 9, n. 2.) It appears that even Nietzsche's and Whorf's thoughts on linguistic determinism are far from unequivocal.

24. Steven Pinker, *The Language Instinct: How the Mind Creates Language* (New York, HarperPerennial, 1994), pp. 59–67. Pinker provides an interesting overview of how these "romantic" notions of cultural relativism were disseminated in American educational institutions from the 1930s onward. From this, I believe, one could partially explain the warm reception that the vulgate version of "postmodernism" has had in American academia. What seems to have emerged is a strangely inconsistent simulation of romanticism. While romantically laboring to refine our sensitivity to other cultures—"[p]ostmodern knowledge [. . .] refines our sensitivity to differences and reinforces our ability to tolerate the incommensurable," claims Lyotard (*The Postmodern Condition*, p. xxv)—postmodern thinkers frequently present culture-in-general in completely unromantic terms, that is, as an instrument of power and torture, something like the horrifying device described in Franz Kafka's story "In the Penal Colony." For example, Deleuze and Guattari (1977) assert that "the movement of culture [. . .] is realized in bodies and inscribed on them, belaboring them. That is what cruelty means. [. . .] It makes men or their organs into parts and wheels of the social machine" (p. 145).

25. See John Searle, *Minds, Brains and Science* (Cambridge, Mass., Harvard University Press, 1984) p. 30. I revisit the thinking thermostat in the next chapter.

26. Let me add, before anyone gets ideas, that I take this example to be utterly wrong in every possible way, and furthermore grossly untrue to Hinduism.

27. Piaget, *Structuralism*, p. 14.

Chapter 9

1. Brouwer, "Mathematics, Science, and Language", in: Mancosu (1998), p. 48.

2. Nietzsche argues that private sensations and individual perceptions cease to be truly individual as soon as they are brought to the level of linguistic description, where they become public and equally accessible to anyone. They thus become "shallow, meager, relatively stupid,—a generalization, a symbol, a characteristic of the herd." More precisely, Nietzsche says that "our thought itself is continuously as it were outvoted by the character of consciousness—by the imperious 'genius of the species' therein—and is translated back into the perspective of the herd. Fundamentally our actions are in an incomparable manner altogether personal, unique and absolutely individual—there is no doubt about it; but as soon as we translate them into consciousness, they do not appear so any longer" (*Complete Works*, Vol. 10, p. 299). Wittgenstein could easily be referring to Nietzsche's ideas about unconscious domination of grammatical functions when he says that philosophy "is a battle against the bewitchment of our intelligence by means of language" (*Philosophical Investigations*, para. 109).

3. Schelling, *Sämtliche Schriften*, 1/10, pp. 11–2. English translation quoted from Frank (1989), p. 299.

4. See Berlin (1999), p. 44.

5. For Schleiermacher, understanding language "has the character of the work of art because [grammatical] rules do not also provide for their application, i.e., their application cannot be mechanized" (*Hermeneutik und Kritik*, p. 81). Quoted from Frank (1989), p. 441.

6. Brouwer, "Mathematics, Science, and Language," in: Mancosu (1998), p. 48.

7. The example is adopted from Saul Kripke, *Wittgenstein on Rules and Private Language* (Boston, Harvard University Press, 1982).

8. Wittgenstein, *Philosophical Investigations*, para. 201.

9. This is the gist of the "Chinese Room" argument against artificial intelligence, which I discuss in detail soon.

10. One finds similar notions in many a "postmodern" reference source. For example, in Hans-Georg Gadamer's *Truth and Method* (New York, Crossroad, 1975), one reads that understanding "is not be thought of so much as an action of one's subjectivity, but as the placing of oneself in the process of tradition" (p. 258). That tradition as a "historical a priori" is a necessary component of understanding is a view that has been in circulation since Hamann and Herder. But if understanding does not involve actions of self-consciousness, tradition would have to be not only a necessary but also a *sufficient* condition for the occurrence of understanding.

11. By the same token, and this is indeed what people who think that thermostats think have argued, an accumulation of roundoff errors in a large-scale computer system would have to be considered the machine's own kind of creativity.

12. Wittgenstein, *Philosophical Investigations*, II-xi, p. 227.

13. "Grammar [. . .] only describes and in no way explains the use of signs" (ibid., para. 496). Schleiermacher's view of grammatical rules, cited in n. 4 above, also seems relevant to Wittgenstein's para. 85: "A rule stands there like a sign-post. [. . .] But where is it said which way I am to follow it; [. . .] it sometimes leaves room for doubt and sometimes not." It is also interesting to compare the "infinite regress" structure of Wittgenstein's private language argument—as well as his statements to the effect that when we ask for reasons for our actions we eventually "run out of reasons" and then simply *act*, "without reason"—with some of Fichte's ideas. Berlin (1999), for instance, describes Fichte's contribution to romantic thought in the following way. "[Fichte] says: if you are simply a contemplative being and ask for the answers to questions such as what to do [. . .], you will never discover an answer. [. . .] Knowledge always presupposes larger knowledge: you arrive at a proposition and you ask for the authority for it, and then some other knowledge, some other proposition, is brought in in order to validate the first one. Then that proposition in turn needs validation [. . .] and so on *ad infinitum*. Therefore there is no end to this search and we simply end up with a Spinozist system, which at best is simply a rigid, logical unity in which there is no room for movement. This is not true, says Fichte. [. . .] 'We do not act because we know,' he says, 'we know because we are called upon to act'" (pp. 88–9).

14. See Rota (1997): "Rumor has it that when Ludwig Wittgenstein died, a worn and marked-up copy of Heidegger's *Being and Time* was found in his quarters at Cambridge University. The writings of the later Wittgenstein arouse the suspicion that the author's purpose was to provide examples for Heidegger's example-free phenomenology" (p. 254).

15. An interesting parallel with the postmodern notion of "performative subjectivity" can be observed here. The basic idea seems to be that self-identity is not "innate," but constituted through—and should be viewed as—a series of "performances" in various language games. For example, American scholar Judith Butler wrote: "[I]dentity is performatively constituted by the very 'expressions' that are said to be its results." (*Gender Trouble*, New York, Routledge, 1990, p. 25.) One might call it a self-referential application of Turing's test.

16. "If the untrained infant's mind is to become an intelligent one, it must acquire both discipline and initiative. *So far we have been considering only discipline* [i.e., Turing machine modeling]. But discipline is certainly not enough in itself to produce intelligence. That which is required in addition we call initiative. This statement will have to serve as a definition. Our task is to discover the nature of this residue as it occurs in man, and to try to *copy* it in machines" (Alan Turing, "Intelligent Machinery," *Machine Intelligence*, **5** [1948], 3–23, p. 21, italics and comment added). Gödel's comments on Turing may also be interesting in this context: "What Turing disregards completely is the fact that mind, in

its use, is not static, but constantly developing" (Hao Wang, *From Mathematics to Philosophy* [New York, Humanities Press, 1974], p. 325).

17. The relevance of this remark in the context of the artificial intelligence project is that it raises the following simple question: How would an intelligent being (e.g., a cognitive scientist) prove that a machine is intelligent, without implicitly presupposing intelligence (e.g., his or her own) in that deduction? If this cannot be ensured, then one can at best claim that intelligent behavior can be *attributed to* the machine by an intelligent being. This seems to be the essence of the Turing test, at least as I understand it.

18. For details, see John Searle, *Minds, Brains and Science* (Cambridge, Mass., Harvard University Press, 1984), Chap. 2, esp. pp. 32–3.

19. Having suppressed the notion of individual creativity, both Foucault and Searle find themselves invoking some magical "powers" instead. For Foucault, it is "power-in-general." For Searle, it is the "power of convention." But unlike Foucault, Searle has not succumbed to antihumanist mysticism: Conventions are ultimately the result of human activity.

20. A difficulty of a similar kind occurs in postmodern theories of subjectivity. The basic idea seems to be that the ego is formed "in the mirror" (the "gaze of the Other," or some such thing). This presumably means that my identity is "assigned" to me through my being reflected in some mirroring structure—social, cultural, economic, material, linguistic, or whatever. This is undoubtedly a necessary component of self-identification. But it does not seem to suffice. How could I possibly identify anything I see in a "mirror" as being *me*, if I am completely unaware of myself to begin with? A formal proof that this can indeed happen would, it seems, also be the proof that computers can develop (postmodern) egos. For an extensive critical analysis of postmodern views on subjectivity, see Frank (1989).

21. In a polemic against Heidegger's notion of understanding, Sartre says: "But how could there be an understanding which would not in itself be the consciousness of being understanding?" Further on, he adds: "We cannot first suppress the dimension 'consciousness,' not even if it is in order to reestablish it subsequently. Understanding has meaning only if it is consciousness of understanding" (*Being and Nothingness* [New York, Philosophical Library, 1956], pp. 73–4, 85).

22. Interesting objections are raised by cultural relativists. The essence of this critique of artificial intelligence is that the structure of most programming languages derives from a fairly narrow group of Indo-European languages. Thus, the artificial intelligence problem is related to a series of other controversial and difficult questions: the validity of the Sapir-Whorf thesis, linguistic relativism, Chomsky's "universal grammar hypothesis," and so on. For a brief discussion of these issues and further references, see Eco (1995), pp. 330–6.

Chapter 10

1. Jacques Derrida, "Deconstruction and the Other," in: Richard Kearney, *States of Mind* (New York, New York University Press, 1995), pp. 172–3.

2. "Die Grundlagen der Mathematik," *Abhaundlungen aus dem Mathematischen Seminar der Hamburgischen Universität*, **6** (1928), 65–85. English translation quoted from van Heijenoort, *From Frege to Gödel*, p. 475. See also Mancosu (1998), p. 160.

3. Jacques Derrida, *Positions* (Chicago, University of Chicago Press, 1981), p. 35. I think this should be understood as saying that people occasionally employ mathematics purely formally, without a clearly preconceived or uniquely *intended* meaning, and address the question of "meaning" only *later*. Simply said, you write down a formally plausible theory and then worry about how it could be interpreted. Physicist Paul Dirac expressed this idea in no uncertain terms: "The most powerful method of advance that can be suggested at present is to employ all the resources of pure mathematics in attempts to perfect and generalize the mathematical formalism that forms the existing basis of theoretical physics, and *after* each success in this direction, to try to interpret the new mathematical features in terms of physical entities" ("Quantized Singularities in the Electromagnetic Field," *Proc. Roy. Soc. London A*, **133** [1931], 60–72). Thus, our "intuitive" understanding may well be retroactively influenced by the process of mathematical formalization. In a somewhat different way, this has also been argued by Heidegger and Wittgenstein. Wittgenstein, for example, says that "mathematical proofs [. . .] lead us to revise what counts as the domain of the imaginable" (*Philosophical Investigations*, para. 517). For Heidegger, see chap. 4, n. 19.

4. Kearney, (1995), p. 173.

5. Derrida, *Positions*, p. 35.

6. Ibid., pp. 35–6.

7. Jacques Derrida, *Writing and Difference* (Chicago, University of Chicago Press, 1978), p. 279 (italics added).

8. Ibid., p. 289 (italics added).

9. Jacques Lacan, *Écrits: A Selection* (New York, Norton, 1977), p. 153 (italics added).

10. See John Searle, "Reiterating the Differences: A Reply to Derrida," *Glyph*, **1** (1977), 199, 207: "Without this feature of iterability there could not be the possibility of producing an infinite number of sentences with a finite list of elements; and this, as philosophers since Frege have recognized, is one of the crucial features of any language. [. . .] *Any conventional act involves the notion of the repetition of the same*" (italics added). There is a pleasant symmetry in the fact that Searle refers to Frege while Derrida's critique of Searle takes the shape of an elaborate reinvention of Poincaré's objections to logicists and formalists (Frege included).

11. Derrida, *Writing and Difference*, p. 167.

12. Derrida, *Positions*, p. 81.

13. See John Caputo, *Deconstruction in a Nutshell: A Conversation with Jacques Derrida* (New York, Fordham University Press, 1997), p. 28.

14. Derrida, *Positions*, pp. 27–28.

15. For a discussion of Kristeva's (blundered) attempt to deal mathematically with the problem of the continuum—without involving intuitionist arguments—

see Sokal and Bricmout, *Fashionable Nonsense*, chap. 3. Lyotard explicitly mentions intuitionism, but only in what is literally a parenthetical remark, and then avoids discussing it: "I am not discussing here the serious objections leveled against this axiomatic model by intuitionism or by the theorem of non-closure of discursive systems" ("What Is Just? (*Ou Justesse*)," in: Kearney [1995], p. 299). It seems to me self-defeating that Lyotard on the one hand defines the "postmodern condition" as a general "incredulity toward meta-narratives," while on the other hand he invokes the most stubborn meta-narrative of Western culture (i.e., mathematical truth) in order to indicate the limits of formal methodology. If we are incredulous toward mathematics, then mathematics cannot be used to "level serious objections" to anything.

16. Heidegger, *Being and Time*, p. 187. This is supposed to deliver the final blow to "self-centered" romantic idealism. But one can find a similar "subversion" of individual intuition already in Fichte: "It is not the individual as such but the one Life which intuits the objects of the material world" (*Sämtliche Werke*, Vol. 2, p. 614). In fact, romanticism occasionally spoke of the empirical self as "feeling absolutely dependent" on some greater creative force.

17. Jacques Derrida, *Of Grammatology* (Baltimore, Johns Hopkins University Press, 1976), p. 61 (comment added). Derrida sometimes also speaks of Heidegger's "nostalgia" for reappropriating "the origin."

18. Richard Lewontin, *Biology as Ideology* (Concord, Ontario, Anansi Press, 1991), p. 41.

19. Jacques Derrida, *Limited Inc.* (Evanston, Northwestern University Press, 1988), p. 53.

20. "To deconstruct the subject does not mean to deny its existence. There are subjects, 'operations' or 'effects' of subjectivity. This is an incontrovertible fact." Quoted from Kearney (1995), p. 175.

21. Jacques Derrida, *Dissemination* (Chicago, University of Chicago Press, 1981), p. 340.

22. Hilbert, "Neubegründung der Mathematik. Erste Mitteilung," *Abhaundlungen aus dem Mathematischen Seminar der Hamburgischen Universität*, **1** (1922), 157–77. English translation quoted from Mancosu (1998), p. 200.

23. Curiously enough, this so-called "post-Cartesian" identity is acquired in a process familiar from textbooks on Keynsian economics: Borrowing from the "virtual" future initiates the desire to balance the debt. I borrow an illusion of identity—for example, from a number of corporate brands—upon which I am made to pay off the debt. But the identification is never complete. Just like in the economy of endless borrowing, the final settling of accounts never happens. It would be the precise equivalent of death, that is, the collapse of the economic system. (This is one way of understanding Lacan's concept of "death drive.") Since these Keynsian tenets seem to be inescapably inscribed into the "economy" of the mind/body, it is not surprising that there are people who—like the Canadian techno-enthusiast Arthur Kroker—consider their Visa cards to be constitutive parts of their "wired" minds/bodies.

24. See Gary Genosko (ed.), *The Guattari Reader* (Oxford, Blackwell, 1996), pp. 109, 112.

SELECT BIBLIOGRAPHY

Arber, Agnes. 1964. *The Mind and the Eye*. Cambridge: Cambridge University Press.
Arnold, Vladimir. 1990. *Huygens and Barrow, Newton and Hooke*. Boston: Birkhäuser.
Audi, Robert (ed.). 1995. *Cambridge Dictionary of Philosophy*. New York: Cambridge University Press.
Barthes, Ronald. 1979. *Critical Essays*. Evanston, Ill.: Northwestern University Press.
Baudrillard, Jean. 1983. *Simulation*. New York: Columbia University Press.
Bencivenga, Ermanno. 1987. *Kant's Copernican Revolution*. New York: Oxford University Press.
Berlin, Isaiah. 1999. *The Roots of Romanticism*. Princeton, N.J.: Princeton University Press.
Boyer, Carl, and Merzbach, Uta. 1989. *A History of Mathematics*. 2nd rev. ed. New York: Wiley.
Brouwer, L.E.J. 1975. *Collected Works*. Ed. by A. Heyting. Amsterdam: North-Holland.
Cahoone, Lawrence E. (ed.). 1996. *From Modernism to Postmodernism: An Anthology*. Cambridge, Mass.: Blackwell.
Caputo, John D. 1997. *Deconstruction in a Nutshell: A Conversation with Jacques Derrida*. New York: Fordham University Press.
Carroll, John B. (ed.). 1956. *Language, Thought and Reality: Selected Writings of Benjamin Lee Whorf*. Cambridge, Mass.: MIT Press.
Cassirer, Ernst. 1957. *The Philosophy of Symbolic Forms. Vol. 3. The Phenomenology of Knowledge*. New Haven, Ct.: Yale University Press.
Cavaillès, Jean. 1984. *Sur la logique et la théorie de la science*, 4th ed. Paris: Vrin.
Chaitin, Gregory J. 1998. *The Limits of Mathematics: A Course on Information Theory and the Limits of Formal Reasoning*. Singapore: Springer.
Chaitin, Gregory J. 1999. *The Unknowable*. Singapore: Springer.

Copleston, Frederick. 1994. *A History of Philosophy. Vol. 7. From the Post-Kantian Idealists to Marx, Kierkegaard, and Nietzsche.* New York: Doubleday.

Damasio, Antonio R. 1994. *Descartes' Error: Emotion, Reason, and the Human Brain.* New York: Avon.

Deleuze, Gilles, and Guattari, Félix. 1977. *Anti-Oedipus: Capitalism and Schizophrenia.* New York: Viking.

Derrida, Jacques. 1978. *Edmund Husserl's Origins of Geometry: An Introduction.* Stony Brook, N.Y.: Nicholas Hays.

Derrida, Jacques. 1978. *Writing and Difference.* Chicago: University of Chicago Press.

Derrida, Jacques. 1981. *Dissemination.* Chicago: University of Chicago Press.

Derrida, Jacques. 1981. *Positions.* Chicago: University of Chicago Press.

Derrida, Jacques. 1984. *Of Grammatology.* Baltimore: Johns Hopkins University Press.

Derrida, Jacques. 1988. *Limited, Inc.* Evanston, Ill.: Northwestern University Press.

Detlefsen, Michael. 1996. "Philosophy of Mathematics in the 20th Century." In: Shanker, Stuart (ed.). *Philosophy of Science, Logic and Mathematics in the 20th Century.* Routledge History of Philosophy, vol. 9. New York: Routledge. 50–123.

Dieudonné, Jean. 1992. *Mathematics—The Music of Reason.* New York: Springer.

Doherty, Thomas (ed.). 1993. *Postmodernism: A Reader.* New York: Columbia University Press.

Dunning, William V. 1995. *The Roots of Postmodernism.* Englewood Cliffs, N.J.: Prentice-Hall.

Eagleton, Terry. 1996. *The Illusions of Postmodernism.* Oxford: Blackwell.

Eco, Umberto. 1995. *The Search for the Perfect Language.* Oxford: Blackwell.

Feferman, Solomon. 1998. *In the Light of Logic.* New York: Oxford University Press.

Fetisov, A.I. 1986. *Očerki po Evklidovoi i neevklidovoi geometrii.* Moscow: Prosvešenie.

Feyerabend, Paul. 1984. *Wissenschaft als Kunst.* Frankfurt: Suhrkamp.

Foucault, Michel. 1970. *The Order of Things: An Archaeology of the Human Sciences.* New York: Vintage.

Foucault, Michel. 1971. *The Archaeology of Knowledge.* New York: Pantheon.

Frank, Manfred. 1989. *What Is Neostructuralism?* Minneapolis: University of Minnesota Press.

Frank, Manfred. 1993. *Conditio Moderna.* Leipzig: Reclam.

Frank, Thomas, and Weiland, Matt (eds.). 1997. *Commodify Your Dissent.* New York: Norton.

Gadamer, Hans-Georg. 1975. *Truth and Method.* New York: Crossroad.

Genosko, Gary (ed.). 1996. *The Guattari Reader.* Oxford: Blackwell.

George, Alexander. 1994. *Mathematics and Mind.* New York: Oxford University Press.

Goldfarb, Warren. 1988. "Poincaré against the Logicists." In: Kitcher, P., and Aspray, W. (eds.). *History and Philosophy of Modern Mathematics.* Minneapolis: University of Minnesota Press. 61–81.

Greimas, Algridas Julien. 1966. *Structural Semantics*. Lincoln: University of Nebraska Press.
Habermas, Jürgen. 1987. *The Philosophical Discourse of Modernity*. Cambridge, Mass.: MIT Press.
Harris, Roy. 1988. *Language, Saussure and Wittgenstein: How to Play Games with Words*. New York: Routledge.
Harrison, Ross. 1983. *Bentham*. London: Routledge & Kegan Paul.
Hart, W.D. (ed.). 1996. *The Philosophy of Mathematics*. New York: Oxford University Press.
Harvey, David. 1989. *The Condition of Postmodernity: An Enquiry into the Origins of Cultural Change*. Oxford: Blackwell.
Hegel, Georg Wilhelm Friedrich. 1967. *Philosophy of Right*. Oxford: Oxford University Press.
Hegel, Georg Wilhelm Friedrich. 1977. *The Phenomenology of Spirit*. Oxford: Oxford University Press.
Heidegger, Martin. 1962. *Being and Time*. New York: Harper & Row.
Heidegger, Martin. 1975. *Poetry, Language, Thought*. New York: Harper & Row.
Heidegger, Martin. 1977. *Basic Writings*. Ed. by David Farrell Krell. New York: Harper & Row.
Heijenoort, J. 1967. *From Frege to Gödel*. Cambridge, Mass.: Harvard University Press.
Hersch, Jeanne. 1993. *L'Étonnement philosophique: Une Histoire de Philosophie*. Paris: Gallimard.
Hersh, Reuben. 1997. *What Is Mathematics, Really?* New York: Oxford University Press.
Hjemslev, Louis. 1963. *Prolegomena to a Theory of Language*, rev. ed. Madison: University of Wisconsin Press.
Honderich, Ted (ed.). 1995. *The Oxford Companion to Philosophy*. New York: Oxford University Press.
Humboldt, Wilhelm von. 1972. *Linguistic Variability and Intellectual Development*. Philadelphia: University of Pennsylvania Press.
Husserl, Edmund. 1960. *Cartesian Meditations: An Introduction to Phenomenology*. The Hague: Nijhoff.
Husserl, Edmund. 1965. *Phenomenology and the Crisis of Philosophy*. New York: Harper & Row.
Husserl, Edmund. 1970. *The Crisis of European Sciences and Transcendental Phenomenology*. Evanston, Ill.: Northwestern University Press.
Kant, Immanuel. 1952. *Critique of Judgment*. Oxford: Clarendon.
Kant, Immanuel. 1978. *Critique of Pure Reason*, trans. by Norman Kemp Smith. New York: St. Martin's Press.
Kant, Immanuel. 1979. *The Conflict of Faculties*. New York: Abaris.
Kearney, Richard. 1995. *States of Mind: Dialogues with Contemporary Thinkers*. New York: New York University Press.
Keynes, John Maynard. 1972. *The Collected Writings of John Maynard Keynes*, Vol. 10. *Essays in Biography*. London: Macmillan.
Kitcher, Philip. 1984. *The Nature of Mathematical Knowledge*. New York: Oxford University Press.

Kline, Morris. 1980. *Mathematics: The Loss of Certainty*. New York: Oxford University Press.
Klotz, Heinrich. 1994. *Kunst im 20. Jahrhundert: Moderne. Postmoderne, Zweite Moderne*. München: Beck.
Koch, Sigmund (ed.). 1959. *Psychology: A Study of a Science*. vol. 1. New York: McGraw-Hill.
Kolers, Paul. A. 1972. *Aspects of Motion Perception*. Oxford: Pergamon.
Koyré, Alexandre. 1973. *Études d'Histoire de la Pensée Scientifique*. Paris: Gallimard.
Kripke, Saul A. 1982. *Wittgenstein on Rules and Private Language*. Boston: Harvard University Press.
Kristeva, Julia. 1968. *Recherches pour une Sémanalyse*. Paris: Seuil.
Kristeva, Julia. 1974. *La Révolution du Langage Poétique*. Paris: Seuil.
Lacan, Jacques. 1977. *Écrits: A Selection*. New York: Norton.
Lakatos, Imre. 1978. *Mathematics, Science and Epistemology*. Lakatos Philosophical Papers, vol. 2. Cambridge: Cambridge University Press.
Latour, Bruno. 1993. *We Have Never Been Modern*. Cambridge, Mass.: Harvard University Press.
Lechte, John. 1994. *Fifty Contemporary Thinkers*. London: Routlege.
Lévi-Strauss, Claude. 1967. *Structural Anthropology*. New York: Doubleday.
Lévi-Strauss, Claude. 1970. *The Raw and the Cooked*. New York: Harper & Row.
Lévi-Strauss, Claude. 1981. *The Naked Man*. New York: Harper & Row.
Levy, Oscar (ed.). 1909–13. *The Complete Works of Friedrich Nietzsche*. New York: Macmillan.
Lewontin, Richard C. 1991. *Biology as Ideology* (CBC Radio Massey lectures series, 1990). Concord, Ontario: Anansi.
Lyotard, Jean-Francois. 1984. *The Postmodern Condition: A Report on Knowledge*. Minneapolis: University of Minnesota Press.
MacIntyre, Alasdair (ed.). 1972. *Hegel*. Garden City, N.J.: Anchor.
Mancosu, Paolo. 1998. *From Brouwer to Hilbert*. New York: Oxford University Press.
McGowan, John. 1991. *Postmodernism and Its Crisis*. Ithaca, N.Y.: Cornell University Press.
McNeill, William, and Feldman, Karen S. 1998. *Continental Philosophy: An Anthology*. Malden, Mass.: Blackwell.
Miller, Arthur I. 1986. *Imagery in Scientific Thought*. Cambridge, Mass.: MIT Press.
Misner, Charles W., Thorne, Kip S., and Wheeler, John A. 1970. *Gravitation*. San Francisco: Freeman.
Newman, J.R. (ed.). 1956. *The World of Mathematics*. New York: Simon & Schuster.
Nietzsche, Friedrich. 1961. *Thus Spoke Zarathustra*. London: Penguin.
Nietzsche, Friedrich. 1967. *The Will to Power*. New York: Random House.
Nietzsche, Friedrich. 1974. *The Gay Science*. New York: Random House.
Panofsky, Erwin. 1954. *Galileo as a Critic of the Arts*. The Hague: Nijhoff.
Passmore, John. 1966. *A Hundred Years of Philosophy*, rev. ed. New York: Basic.
Peirce, Charles Sanders. 1953. *Collected Papers*. Ed. by C. Hartstone and P. Weiss. Cambridge, Mass.: Harvard University Press.
Piaget, Jean. 1969. *The Psychology of the Child*. New York: Basic.

Piaget, Jean. 1970. *Structuralism*. New York: Harper & Row.
Pinker, Steven. 1995. *The Language Instinct: How the Mind Creates Language*. New York: HarperPerennial.
Poincaré, Henri. 1934–53. *Ouevres de Henri Poincaré*. Paris: Gauthier-Villars.
Poincaré, Henri. 1952. *Science and Method*. New York: Dover.
Poincaré, Henri. 1963. *Mathematics and Science: Last Essays*. New York: Dover.
Propp, Vladimir. 1968. *Morphology of the Folktale*. Austin: University of Texas Press.
Rée, Jonathan. 1998. *Heidegger*. London: Phoenix.
Rota, Gian-Carlo. 1997. *Indiscrete Thoughts*. Boston: Birkhäuser.
Ruelle, David. 1991. *Chance and Chaos*. Princeton, N.J.: Princeton University Press.
Sartre, Jean-Paul. 1956. *Being and Nothingness*. New York: Philosophical Library.
Saussure, Ferdinand de. 1966. *Course in General Linguistics*. New York: McGraw-Hill.
Saussure, Ferdinand de. 1967–74. *Cours de linguistique générale*. Edition Critique. Ed. by Rudolf Engler. Wiesbaden: Harrasowitz.
Searle, John. 1969. *Speech Acts: An Essay in the Philosophy of Language*. Cambridge: Cambridge University Press.
Searle, John. 1984. *Minds, Brains and Science*. Cambridge, Mass.: Harvard University Press.
Šikić, Zvonimir (ed.). 1987. *Novija filozofija matematike*. Beograd: Nolit.
Sloterdijk, Peter. 1987. *Kopernikanische Mobilmachung und Ptolemäische Abrüstung*. Frankfurt: Suhrkamp.
Sokal, Alan, and Bricmont, Jean. 1998. *Fashionable Nonsense: Postmodern Intellectuals' Abuse of Science*. New York: Picador.
Stewart, Ian. 1992. *The Problems of Mathematics*, 2nd ed. Oxford: Oxford University Press.
Struik, Dirk J. 1987. *A Concise History of Mathematics*. New York: Dover.
Trostnikov, V.N. 1975. *Konstruktivnye processy v matematike*. Moscow: Nauka.
Tymoczko, Thomas (ed.). 1998. *New Directions in the Philosophy of Mathematics*, 2nd rev. ed. Princeton, N.J.: Princeton University Press.
Van Dalen, D. (ed.). 1981. *Brouwer's Cambridge Lectures on Intuitionism*. Cambridge: Cambridge University Press.
Vattimo, Gianni. 1991. *The End of Modernity: Nihilism and Hermeneutics in Postmodern Culture*. Baltimore: Johns Hopkins University Press.
Wang, Hao. 1974. *From Mathematics to Philosophy*. New York: Humanities.
Weyl, Hermann. 1963. *Philosophy of Mathematics and Natural Science*. New York: Atheneum.
Wittgenstein, Ludwig. 1953. *Philosophical Investigations*. Oxford: Blackwell.
Wittgenstein, Ludwig. 1958. *The Blue and Brown Books*. Oxford: Blackwell.
Žižek, Slavoj. 1993. *Tarrying with the Negative: Kant, Hegel, and the Critique of Ideology*. Durham, N.C.: Duke University Press.

INDEX

Adams, Douglas 78
Adian, Sergey 28
artificial intelligence 112, 134, 174 n. 17
   and Lacan 168 n. 14
   and postmodern subject 95–6, 168 n. 14, 174 n. 20
   and Sartre 136–7
   and Schelling 136
   thesis, strong form of 135–6
   thesis, weak form of 137

Barthes, Roland 113
Baudrillard, Jean 165 n. 20
Becker, Oskar 166 n. 21
Bentham, Jeremy 26–7, 160 n. 9
Bergson, Henri Louis 34, 37–40, 42–3, 85, 112, 167 n. 1
Berlin, Isaiah 34, 173 n. 13
Blake, William 35
Bohr, Niels 88
Bolyai, Janos 21–2, 24
Bombelli, Rafael 87
Borel, Emile 52, 81–2
Borges, Jorge Luis 41
Bourbaki, Nicholas 112
Boyer, Carl 30

Brouwer, Luitzgen Egbertus Jan 36–50, 54, 56, 58, 61, 63, 68–70, 74, 89, 94, 107, 147–50, 152, 154, 162 n. 19
   on certainty and language 46, 48–9
   and Heidegger 42–5
   on intuition of time 37
   and language as ideology 47–8
   on mathematical attention 47
   on mathematics and language 46
   and Nietzsche 46–7
   and Principle of the Excluded Middle 40–1
   on science and logic 45–6
   on the ultimate philosophical plug 45
   on will-transmission 46
   and Wittgenstein 49, 119–20, 123, 126, 130
Butler, Judith
   on peformative subjectivity 173 n. 15
Byron, Gordon George 5, 97

Canguilhem, Georges 86, 90
Cardano, Girolamo 87
Carrington, Leonora 160 n. 10

Cartesian Circuit 8, 10, 17, 153
Cavaillès, Jean 73, 84–6, 89, 90, 92–3, 111
   as a critic of Kant and Husserl 85
   and formalism 85
   and Gödel 89
   rebels against romanticism 100, 167 n. 1
   and structuralism 100, 167 n. 1
center 143
Chaitin, Gregory 79–83, 91, 97 (see also randomness; halting probability)
Chinese Room argument 134–5, 172 n. 9
choice-sequence 39–40, 147
Chomsky, Noam
   and models of language 79, 95, 102
   and Piaget 153, 165 n. 18
Church, Alonzo 77
Church-Turing thesis 77–9
Coleridge, Samuel Taylor 32
constructivism 38–9
continuum 36–43
   as compared to *différance* 146–150
   constructions of 39–40, 53, 62
   and Heidegger 42–4
   and Husserl 55
   and incompleteness 82
   as the "other of language" 50, 52, 56
   and schizophrenia 98
   and undecidability 51
counter-enlightenment 5
culture (see also determinism)
   as informative of reasoning 28–9
   and mathematics 28, 90

Damasio, Antonio 59
Darwin, Charles 87
deconstruction 4, 141, 145, 157 (see also Derrida, Jacques)
Dedekind, Richard 106–8
   and structuralist ideas 106

Deleuze, Gilles 98–9
   and management literature 169 n. 16
   on Oedipal and anti-Oedipal thought 98
   on schizophrenic logic 98
Derrida, Jacques 38, 62, 138–54, 157
   on centered structures 143
   on *différance* 147, 149
   on the end of the book 140
   on excesses of formalization 141–2
   on the finitude of language 144
   and formalism 140–2
   and Foucault 97
   on grammatology 142
   and Heidegger 57–8, 150–1
   and Husserl 57–8, 89
   on iterability 146, 152, 175 n. 10
   language in the work of 139
   on mathematics and deconstruction 141
   on the origin 151
   and Poincaré 145–6, 175 n. 10
   and Searle 146, 175 n. 10
   on subjects 153, 176 n. 20
Descartes, Réne 7–10, 23, 33–4, 121
determinism
   cultural 115–6
   linguistic 115, 171 n. 23, 172 n. 2
   and postmodernism 156
   technological 115
Dieudonné, Jean 31
différance 147–53
   as becoming in general 147
   as compared with the continuum 146–50
   as "the mathematical" 150
   as "the semiotic" 149
Dirac, Paul 175 n. 3
discontinuity
   and Foucault 91, 93, 96–7
   and postmodernism 36
discursive practices 91

Eco, Umberto 161 n. 12
Einstein, Albert 10, 16, 22, 53
Euclid 20, 60
Euclidean geometry (*see* geometry)

Ferro, Scipione del 87
Fichte, Johann Gottlieb 15–6, 31, 104, 120, 153
 on I and not-I 15–6, 123
 on intuition 176 n. 16
 on knowledge and action 16
 and limits of introspection 161 n. 17
 and Poincaré 164 n. 14
 and the self 16
 and Wittgenstein 130, 173 n. 13
finitary intuition 68–9, 72–3, 76–7
 and the Church-Turing thesis 79
 as intuition of signs 68
 modeled as Turing machine 78
 as a precondition for logical inferences 68
formalism
 and Cavaillès 84–9
 and Derrida 140–2
 and Foucault 90–6
 Hilbert's school of 67–74
 and structural anthropology 109–11
Foucault, Michel 73, 84, 86, 88, 90–8
 archeology in the sense of 96
 on authorship 95
 and Chaitin's theorem 97
 continuity rejected by 96
 on Deleuze 98
 and Derrida 97
 on disappearance of man 94, 168 n. 12
 and formalism 90–6
 on human sciences 92–4
 "intuitionist" objections to 94
 Piaget on 168 n. 15
 reference to mathematics by 95
 Sartre on 96–7

Frege, Gottlob 38, 44, 58, 119
 invoked by Searle 175 n. 10
 and Principle of Contextuality 105
Fukuyama, Francis 88
functionalism 113, 137–9, 156
 and impredicative definitions 118
 in literary theory 114
 in theories of language and culture 115–6
 in the theory of mind 116–7

Gadamer, Hans-Georg
 and prejudice 61
 on subjectivity 172 n. 10
Galileo 4, 23–4, 26, 55
Gauss, Carl Friedrich 21–2
geometry 20–3, 30–1, 160 n. 10
 Euclidean 21–2
 Hilbert and the foundations of 67–8
 Kant's statements regarding 20
 non-Euclidean 21–2, 59
 Poincaré and the foundations of 60–1
 of space 23–4
Gödel, Kurt 13, 77, 82
 and critique of formalism 74, 111
 the incompleteness theorems of 74–6
 and Leibniz 74
 on mathematical truth 76
 on minds and machines 173 n. 16
 philosophical views of 76
 as a reader of Kant and Husserl 76
 on reasons for incompleteness 89
Goethe, Johan Wolfgang von 16, 69, 84
Goldfarb, Warren
 on Poincaré 65
grammatology 142
Greimas, Algridas Julien 115
Guattari, Pierre-Félix
 on postmodernism 157
 on schizophrenic thought 98

Index 185

halting probability 80–3
Halting problem
  undecidability of 51–2, 80–1
Hamann, Johann Georg 29–30, 102, 121
Hegel, Georg Wilhelm Friedrich 17–9, 24, 36, 67, 84–5, 87–9, 104, 112, 136, 167 n. 2
Heidegger, Martin 4, 42–3, 56–7, 61–2, 92–3, 131, 147, 149–52, 173 n. 14, 174 n. 21
  and Brouwer 42–5
  on deriving intuition 151
  and future-directedness 42–3
  on intuition and formalization 162 n. 19, 175 n. 3
  on intuitionism and formalism 163 n. 20
  on logic 43
  on "the mathematical" 43
  on method 167 n. 6
  on Newton 169 n.1
  subject as defined by 42, 150–1
  on truth in art 44
Helmholz, Hermann von 22
Heraclitus 17, 19, 57, 104
Herder, Johann Gottfried von 28–30, 102, 112, 115
hermeneutical circle 33
hermeneutics 61
Hilbert, David 66–7, 94, 96, 101, 106–7, 109–11, 130, 140–1
  and cognition as computation 77–9
  and finitary intuition 68–9
  on geometry 67–8
  Gödel's comment on 74
  Hegelian reading of 84–7
  on intuition and formalization 141
  on the intuitionist revolution 85, 154
  and Kant's epistemology 68, 70
  and language games 70, 166 n. 7
  and Nelson 72–3
Hitchcock, Alfred 36

Hjemslev, Louis
  on the algebra of language 170 n. 9
Humboldt, Wilhelm von 35, 47, 104, 106, 122
Hume, David 8–10
Husserl, Edmund 4, 44, 53–8, 61, 63, 70, 74, 76, 85, 89
  Derrida's critique of 57, 89
  on loss of meaning 72, 166 n. 6
  on time-consciousness 55–7
  and Weyl 53

ideal mathematics 69–71, 109
ideal elements (method of) 86–7
identity 60–6, 145–6, 152, 164 n. 14
imaginary numbers 87–8
impredicative definitions 63–5, 118, 142–5, 154
incompleteness theorems 74–6, 89
intuitionism 41, 46, 69, 93, 99, 112–3, 139, 142, 146, 148–9, 154, 176 n. 10
iterability 146, 152, 175 n. 10

Kafka, Franz 171 n. 24
Kant, Immanuel 10–5, 24, 26, 28–30, 36, 38, 57, 71, 76
  and antinomies of pure reason 12, 23
  on concepts and experience 10
  on the conflict of faculties 30
  on different logics 13
  on divisibility 40
  and geometry 20–2
  and Hilbert 67–70, 84–5, 87–8, 93, 166 n. 4
  on individual and community 15
  on intuiting ourselves 14
  and science wars 20, 30
  and transcendental illusions 25
Keynes, John Maynard 8, 176 n. 23
Klein, Felix 60
Kristeva, Julia 141, 149, 175 n. 15
Kroker, Arthur 176 n. 23

Lacan, Jacques
    on cameras and machines 168
        n. 14
    and the economy of mind 176
        n. 23
    on meaning 145
    and the problem of induction 166
        n. 21
Lakatos, Imre 159 n. 1
Latour, Bruno 4
Leibniz, Gottfried Wilhelm 7, 13, 27,
        72, 87, 161 n. 12
    and blind thought 27, 72
    and Gödel 74–6
    pretends 28
Lévi-Strauss, Claude 108–12
    and formalism 109–11
    on the logic of myths 110
    and the myth of mathematical
        truth 109–11
    and mythical time 110
    on the operation of myths 111
Lewontin, Richard 151
Lobachevsky, Nikolai Ivanovich 21,
    24, 60
logical positivism 27
Lyotard, Jean-Francois
    on dispersion of language games
        167 n. 7
    invokes intuitionism 176 n. 10
    on local determinism 168 n. 12
    on postmodern knowledge 169
        n. 16

Mao Tse-Tung (Chairman Mao)
    116
Marx, Karl 116, 171 n. 22
McCarthy, John 117, 137
McLuhan, Marshall 116
Merleau-Ponty, Maurice 63
meta-mathematics 71–2
Monterroso, Augusto 56

Nelson, Leonard 72–3, 167 n. 8
Newton, Isaac 8, 169 n. 16,
    169 n. 1

Nicholas of Cusa 23
Nietzsche, Friedrich Wilhelm 34, 49,
        53, 97, 131
    and Brouwer 46–7
    on grammatical functions 114,
        116, 171 n. 23
    and linguistic determinism 116,
        171 n. 23, 172 n. 2
    on logic 62
    and the private language
        argument 172 n. 2
    and the prison-house of language
        97, 171 n. 23, 172 n. 2
    on the universe 34
    and Wittgenstein 172 n. 2
Novalis
    on philosophy 35
    on poetry 32
    on reflection 16

Ortega y Gasset, José 119

paradox 58, 107
    Berry's 142
    of the liar 74
    of the ravens 24–5
    Wittgenstein's 126
    Zeno's 42, 50
Peirce, Charles Sanders 104–5,
    150, 162 n. 13
Peters, Tom 169 n. 16
phenomenology 53–8, 151 (*see also*
        Husserl, Edmund)
    Derrida's critique of 57–8
    Hilbert and the fate of 73
Piaget, Jean 153
    and Bourbaki 112
    and Chomsky 165 n. 18
    on Foucault 169 n. 15
    and Poincaré 165 n. 18
    on self-regulating structures
        118
Pinker, Steven 116, 171 n. 24
Plato 4, 164 n. 8
Platonism (Platonist) 70, 76, 156
Poe, Edgar Allan 100

Poincaré, Jules Henri 50, 58–67, 106–7, 118, 142, 165 n. 21
   and arguments of Fichte and Schelling 164 n. 14
   on continuity 59–60, 62, 165 n. 18
   and Derrida 145–6, 175 n. 10
   on geometry 60–2
   on identity 60–6, 145–6
   and impredicative definitions 63–5, 118, 138, 142–5
   on *logistique* 58, 164 n. 8
   and prejudice 61
   and Saussure 65
Post, Emil 77
postmodernism 4–5
   and chaos theory 155–6
   as functionalism 156
   Guattari on 157
   popular forms of 154
   and romanticism 29, 155
   and Schiller 154–6
   as an umbrella term 4
poststructuralism 4, 97, 118, 138–9, 143
Principle of the Excluded Middle
   and binary thinking 41, 98
   Brouwer on 40–1
   and schizophrenia 98
private language argument 120–30
   and Brouwer 120–2
   and Chinese Room argument 135
   different interpretations of 129–30
   Fichte and the structure of 173 n. 13
   and Nietzsche 120, 172 n. 2
   and rule-following 129
Propp, Vladimir 114
Pythagoras 18, 109

randomness 52, 79–83, 91
real mathematics 69, 71
romanticism 5–6, 16–7, 29–36, 102, 106, 115
   confronted with science 26
   mystical and religious roots of 163 n. 30
   and nostalgia 35
   as opposed to romantic idealism 32
   and paranoia 36
   and postmodernism 29, 155
   on understanding and art 32
   and Wittgenstein 121–2, 130–1, 173 n. 13
Rota, Gian-Carlo
   on Heidegger and Wittgenstein 173 n. 14
   on mathematics and philosophy 113
Russell, Bertrand 4, 38, 44, 58, 70, 74, 90, 107, 119

Saccheri, Girolamo 21
Sapir-Whorf Thesis (*see* Whorf, Benjamin Lee; determinism)
Sartre, Jean-Paul 36, 85, 167 n. 1
   and artificial intelligence 137
   on Foucault 96
   and Heidegger 174 n. 21
   on understanding 174 n. 21
Saussure, Ferdinand de 63, 65, 100–8, 111–2
   on language and community 108, 170 n. 12
   on language and individual 108
   on language and thought 103
   and mathematics 105–7
   and Schleiermacher 106–7
   and structuralism 100–8
Schelling, Friedrich Wilhelm Joseph 40, 161 n. 17, 164 n. 14
   and artificial intelligence 136–7
   on Descartes 34, 121
   on Hegel 136
   and Poincaré 164 n. 14
Schiller, Johann Christoph Friedrich von 154–6
Schlegel, Friedrich von 34–5, 122
Schleiermacher, Friedrich 33, 35, 47, 122, 131
   on language and thought 104, 169 n. 5
   on rules 172 n. 5

and structuralism 106–7
on the work of art 33, 121, 172 n. 5
Schopenhauer, Arthur 36
science wars 3, 5, 157
Searle, John 174 n. 19
  and the Chinese Room 134–7
  Derrida's confrontation with 146, 175 n. 10
self-consciousness
  and cognition 137
  immediate 136
  "in general" 16, 161 n. 17
  and language 16, 135
  as a precondition of understanding 136
  and prelinguistic familiarity 33–4
  romanticist view of 33–4
Skinner, B.F. 129
Smiley, Jane 169 n. 16
Spinoza, Benedict 4, 173 n. 13
structuralism 100–13, 118, 139
  and linguistics 100–5
  and mathematics 105–13
  as opposed to functionalism 113
  as rebellion against romanticism 167 n. 1
subjectivity
  and corporate brands 154
  Heidegger's derivation of 42, 150–1
  performative 173 n. 15
  postmodern 95–6, 168 n. 14, 174 n. 20
sublation (*Aufhebung*) 18–9

Tarski, Alfred 76, 88
Tartaglia, Niccolo 87
Tasso, Torquato 26
Thom, Réne 113
time-consciousness 55–6
time continuum (*see* continuum)
Turing, Alan 82
  and the Church-Turing thesis 77–9
  on discipline and initiative 133, 173 n. 16
  and Halting problem 51–2
  and Wittgenstein 132–3
  Turing machine 77–9, 91, 136
  Turing's test 132–3, 173 n. 15, 174 n. 17

undecidability (undecidable)
  in Derrida's sense 145
  in the mathematical sense 52

Vico, Giambattista 9–11, 17, 28, 39

Weeks Jeffrey 23
Weyl, Hermann 50–8, 63, 70, 73–4, 89, 93, 106, 141, 147
  and Brouwer 53–4, 168 n. 9
  on the continuum 53–4
  and Husserl 53–5
  and structuralist ideas 105
Whitehead, Alfred North 38, 44, 90
Whorf, Benjamin Lee 115, 171 n. 23
  (*see also* determinism)
Wittgenstein, Ludwig 4, 28, 42, 47, 54, 118–23, 126–8, 130–5
  and Brouwer 49, 119–20, 122–3, 126, 130
  and the Chinese Room argument 135
  and Fichte 130, 173 n. 13
  and formalism 70, 130
  and Hamann 121
  and Heidegger 131, 173 n. 14
  on interpreting unknown symbols 121
  and Nietzsche 172 n. 2
  romanticism in the thought of 121–3, 130–1, 173 n. 13
  and Schelling 121
  and Schleiermacher 121, 131, 173 n. 13
  on self-identity 61
  and Turing 132–3

Žižek, Slavoj 160 n. 9
Zweig, Stefan 37